A CD-ROM accompanies this book.
Both items must be returned in order to
discharged from your card.
Any late items are subject to fines.

D0335458

7 Day Loan

ronment

This book is due for return on or before the last date shown
below

2 2 OCT 2007		
1 4 IAN 2008		

St Martin's College

Introducing Environment

Alice Peasgood and Mark Goodwin

Published by Oxford University Press, Great Clarendon Street, Oxford OX2 6DP
in association with The Open University, Walton Hall, Milton Keynes MK7 6AA.

  OXFORD
UNIVERSITY PRESS

Oxford University Press is a department of the University of Oxford. It furthers the
University's objective of excellence in research, scholarship, and education by publishing
worldwide in

Oxford New York Auckland Bangkok Buenos Aires Cape Town Chennai
Dar es Salaam Delhi Hong Kong Istanbul Karachi Kolkata Kuala Lumpur Madrid
Melbourne Mexico City Nairobi São Paulo Shanghai Taipei Tokyo Toronto

Oxford is a registered trade mark of Oxford University Press in the UK and in certain other
countries.

Published in the United States by Oxford University Press Inc., New York

First published 2007

Edited, designed and typeset by The Open University

Printed and bound in the United Kingdom by the Alden Group, Oxford

This book forms part of the Open University course Y161 Introducing Environment. Details of this
and other Open University courses can be obtained from the Student Registration and Enquiry Service,
The Open University, PO Box 197, Milton Keynes MK7 6BJ, United Kingdom:
tel. +44 (0)870 333 4340, email general-enquiries@open.ac.uk.

http://www.open.ac.uk

British Library Cataloguing in Publication Data available on request

Library of Congress Cataloging in Publication Data available on request

ISBN 9780 1992 1713 7

10 9 8 7 6 5 4 3 2 1

Contents

Preface

This book is for anyone with an interest in the environment, but no previous experience of studying science. It does not claim to be comprehensive, but provides an introductory guided tour of some significant scientific and technological ideas. There is hardly any mathematical content – the emphasis is on interpreting charts and graphs rather than calculations. New concepts and specialist words are explained with examples and illustrations. Study skills, such as planning your time and making notes, are developed along the way. Numerous activities, and study checklists at the end of each chapter, give you a chance to learn actively, trying out your understanding as you progress. Chapter 1 explains how the activities, information boxes and other features of the text can support your learning. By the end of the book, you should be better equipped to make sense of environmental issues, and to write about them.

The accompanying DVD* contains over two hours of audio and video, including real-life examples and interviews with experts. In the UK, all you need is a DVD player with the standard menu controls. Activities in the text will refer you to the DVD at appropriate points.

Although written with Open University students in mind, this book is suitable for beginners in ecology, environmental science or related subjects, or general readers who want to go into more depth than is available from news reports. Secondary school teachers and other staff supporting the new 'Education for Sustainable Development' approach in the National Curriculum should find the content relevant as background reading. In such a wide-ranging subject, we have had to be selective, although we have included different aspects of the environment: biology, technology, built environment, sustainable development and ecological footprint. The contents list shows the range of topics covered. Our intention is to bring the novice reader into the subject, develop confidence through activities and advice, and provide a stepping stone into further study.

Authors' acknowledgements

As ever in The Open University, this book combines the efforts of many people with specialist skills and knowledge in different disciplines. It would be impossible to thank everyone personally, but we would like to acknowledge the help and support of colleagues who have contributed to the production of this book.

This could not have happened without the course team who planned the content, discussed drafts, contributed information and generally ensured that all the diverse issues were thoroughly considered: Penny Parkinson (Centre for Widening Participation), Carlton Wood (Science Faculty), James McLannahan (Technology Faculty), Sally O'Brien (Course Team Secretary), George Marsh (Openings Programme Manager). Throughout, Jim Bailey, as Course Manager, provided all the patience, humour and project management skills required to keep the book and DVD on track. Within the Centre for Widening Participation, we would also like to thank Christine Wise (Director), Kate Levers and Pat Spoors (Assistant Directors) for their valuable help at critical moments.

* The DVD is an audio/video multi-region DVD in PAL format which is not compatible with NTSC-only DVD players, although it can be played in many computer DVD drives.

We are very grateful to critical readers from elsewhere in the OU, whose expertise has significantly shaped the content of the book: Jeff Thomas (Science), Maggie King (Technology), Phil Sarre (Social Sciences). In addition, we received helpful comments from student testers in various parts of the UK. We have benefited considerably from the insights of our External Assessor, Edgar Jenkins, Emeritus Professor, School of Education, University of Leeds. His contributions from the perspective of the public understanding of science have set the content within a wider context.

Special thanks are due to all those involved in the OU production process, including Helen Sturgess (Media Project Manager), Debbie Crouch (Graphic Designer), Vicky Eves (Graphic Artist), Kirsten Brown (Media Assistant), Kathy Eason (Production Editor), Bob Heasman (Head of Production Support), Siggy Martin (Assistant Print Buyer). Our editor, Clare Butler, has applied thoughtful professionalism that has balanced the challenging issues in the text.

This book has built on best practice from previous OU courses, and we would like to acknowledge some recycling of text and illustrations from other OU publications, including:

Our Living Environment (1999) Jeff Thomas, The Open University (K507/6)

The Sciences Good Study Guide (1997) Andrew Northedge, Jeff Thomas, Andrew Lane and Alice Peasgood, The Open University.

For the new DVD video material, we would like to thank Chris Chappell and colleagues at Autonomy Multimedia for their dedicated work; and The Co-operative Group for generously allowing recording at their suppliers and in their supermarkets. Within the OU, special thanks to Nicholas Watson (Head of Sound and Vision) and Liaket Ali (Media Production Technician). We would like to acknowledge re-use of OU audio and video material originally produced for K507 *Our Living Environment* and T172 *Working with our Environment: Technology for a Sustainable Future*.

For the copublication process, we would especially like to thank Jonathan Crowe of Oxford University Press and, from within The Open University, Giles Clark (Copublishing Advisor) and Christianne Bailey (Media Developer, Copublishing).

As is the custom, any errors remain the responsibility of the authors. Please send any constructive comments to Alice Peasgood at the address below.

Alice Peasgood (Y161 Course Team Chair) and Mark Goodwin,
Centre for Widening Participation, The Open University, Walton Hall, Milton Keynes
MK7 6AA, United Kingdom

1 Our environment

1.1 Introduction

On 14 February 1990 the space probe Voyager 1 reached the outer edge of the solar system. Before it was lost to deep space, the astronomer Carl Sagan asked the engineers to turn the cameras around for one last image of home. Sagan described the Earth in the resulting photograph, viewed from a distance of almost four billion miles, as a 'pale blue dot'.

That 'pale blue dot' is home to us and to all of the rest of the living things that we know about. There may be life elsewhere in our solar system, or in our galaxy, or in the universe as a whole … but we haven't found it.

Life is everywhere on the Earth, from the poles to the equator, and from the atmosphere above Mount Everest to the bottom of the deepest ocean trench. This life-supporting layer is so thin that it cannot be seen viewed edge on from space – and if the Earth were the size of a football, this fragile layer of life would be as thick as a single coat of paint.

But this thin film of life is astonishingly beautiful, diverse and complex. The interrelationships between plants, animals, oceans, land and atmosphere have developed over billions of years. Many, perhaps most, of these connections are not fully understood, and the features we do understand are amazingly complicated and intricate. It isn't possible to cover such a huge topic in a short book, but the aim is to provide an overview of some of the main pieces of the puzzle. This book is intended for non-specialist readers, so the scientific ideas will be introduced as you go along. There are hardly any calculations – and the ones that are needed are kept simple – so the focus is on comparing numbers and amounts, rather than complicated mathematics.

Figure 1 The Earth from space

Allow about 5 minutes

What does 'environment' mean to you?

Write a few words or phrases that describe what you understand by 'environment'.

Comment

'Environment' means different things to different people. Here are some possibilities; you may have thought of others:

- wildlife and vegetation
- conservation and recycling
- being 'green'
- a place to live
- global warming and climate.

The term **environment** means 'surroundings'. However, different things have different surroundings, so when we talk about 'the environment' we need to be clear what we mean. For example, where does my environment end? My immediate environment is quite small, perhaps my home itself and a few square miles around it. But I am familiar with a much wider area. I travel within the UK to visit family and friends, and occasionally take holidays abroad. The food and other resources I use come from even further afield. On the largest scale, I am one of all living things that share the same environment: the biosphere itself. The **biosphere** includes all life on Earth, as well as the parts of the planet that support life – rocks, soils, waters and atmosphere. The study of the biosphere is therefore the study of life on Earth. The biosphere forms a thin but seamless layer around the surface of the Earth.

The environment of any living thing consists of other living things and a collection of non-living resources (air, water, soil and so on). It is necessary to consider both the living and the non-living components as the two are closely related and interdependent. Life evolved on Earth somewhere between 3.5 and 4.5 billion years ago. It has been shaped by – and has shaped – the land, water and atmosphere ever since. Living things shape their environments, and are shaped by the environments in which they live. This is an important point because changes to an environment can have different consequences for the different living things that share the living space. Draining a ditch is bad news for mosquitoes, which need water in which to lay their eggs, but good news for the human beings and other animals that would otherwise be bitten by the next generation of adult insects.

In this chapter, you'll be introduced to three overarching ideas that relate to everything in this book: environment, science and technology. As you have seen, the first time a key word is explained in the main text, the word is highlighted in green. You have already encountered the first activity, and there are many more throughout the text. This book is written to encourage active learning, so the next section will explain what this means, and give you a chance to tackle some specialist reading about the age of life on Earth. The purpose of the text boxes and checklists will be explained, and there will be a brief tour of the contents of the rest of the book. Finally, you will have a chance to review your study of this chapter, and plan for Chapter 2, where the main environmental story continues.

1.2 Active learning

This book is designed for active learning, so there are places in the text where you are asked to stop reading and do an activity, write down your thoughts or use the audio and video material on the DVD. This section describes how to tackle these activities.

The first step is to get organised – find somewhere comfortable to study where you can set out your materials to work on, and then gather together the extra things you need – pens, pencils, eraser, paper, DVD player, and so on. Even if you are studying this book more casually, you will find it useful to have at least a pen or pencil and some extra paper for notes. A small notebook is particularly useful, as you can keep your jottings in order, and note any queries as you go along. Or you may prefer to write your notes on paper and store them in a folder. All your notes are then kept together and you can look back over your work as a whole.

Occasionally, there are information boxes, such as the one below. These summarise specific advice or information about a topic.

Writing on the book

You can write on the book by highlighting the important information in the text, assuming the book is yours to keep. The margins have been deliberately left wide, to give you space for notes. You might also like to write comments to remind yourself of questions you would like to clarify in future. You may like to use coloured pens to highlight different aspects of the text; for example, blue for important definitions, yellow for useful examples, pink for sections relevant to any course work or assignments, and so on. This will make things easier to find if you need to read back over your work later. Some people prefer to use a pencil, so they can change their comments as their understanding of the subject alters.

For some people, the activities and instructions can seem irritating – why interrupt my reading? – but the aim is to help you to think about the subject, and to engage more deeply with the content of the book. It is assumed that you will do each activity at the relevant point in the text. You'll find that the discussion continues after the comment. For some activities, where the response would be too obvious printed directly below the question, the comment is at the back of the book – you will be referred to these at the relevant places.

For each task, there is a suggested time, for example, 'Allow 10 minutes'. These estimates are intended to give you a sense of the amount of effort required. If the subject is unfamiliar, you may find that you spend longer on each activity. That's fine, so long as you feel you are learning. If you come across ideas that you have encountered before, you may work through the activities more quickly. Effective learning does not have to take hours at a stretch; in fact, trying to concentrate for too long can be less efficient as you become tired. You are the only one who can tell what works best for you. At first, it may be worth noting the actual time spent on each task, so you can decide whether to change the way you study.

Some tasks introduce new information, to give you practice in dealing with unfamiliar words or ideas. Others may refer back to earlier sections of the book, so you can consider how new ideas link with ones discussed previously. The next activity introduces you to three different writing styles. It assumes that the words and ideas are new to you. Although new terms appear in green in the main text, this is not the case for quoted text, so it is up to you to underline or highlight any words that you think are important. What matters is how you tackle this task – even if you find it a challenge. You may want to make notes or draw a diagram as you read, and you will probably need to read the examples more than once – this is normal for this type of text, even for the most experienced reader.

Activity 2

Allow about 30 minutes

Reading text – the age of life on Earth

The three examples of text in this activity come from three different sources: Example 1 is from a popular science magazine, Example 2 is from a student textbook for a higher level than Introducing Environment, and Example 3 is from a book intended for the general reader. Note that a billion is a thousand million, and a million is a thousand thousand. These three texts are probably more complicated than anything else in this book, so the comment afterwards will give you some advice about how to tackle complex and unfamiliar text. After this activity, the rest of the book should be less daunting. You don't need to remember the information in these examples – what matters more is how you approach the reading.

Read each of the examples, then note:

- how they compare – which is the hardest to understand, which is the most 'scientific'
- how well you can make sense of them – whether particular words are difficult, or whether the numbers are a challenge
- briefly, what the examples are about – jot down a few words or phrases.

Example 1

From all that they can find recorded worldwide in the rocks, geologists have built up a timescale, setting in sequence all the major events that have happened to our planet: it is the keystone to the Earth's 4.6 billion years of evolution.

Geologists can trace the history of the Earth back about 4.6 billion years, to its formation from a ring of gas and dust around the young Sun. They divide this vast span into intervals that form the basic yardsticks of geological time. Early geologists named these intervals on the basis of the rocks formed within them but without knowing how long they lasted. Succeeding generations have changed the names of some and calibrated them in years to produce a geological time scale – a means of measuring the history of the Earth.

(Hecht, 1995)

Example 2

One of the most important events in the history of life began about 545 million years ago, i.e. some four billion years after the origin of the Earth. The term Cambrian explosion reflects a sudden burst of evolution, when a wide variety of organisms, especially those with hard, mineralised parts, first appear in the fossil record. Thus began the Phanerozoic Eon – 'the time of visible life'. Very small (1–2 mm) shelly fossils appeared in the earliest part of the Cambrian Period – assorted shapes such as tubes and cones (that presumably enclosed soft tissue), as well as spines, scales and knobs. It's often difficult to tell, however, whether a fossil is the complete skeleton of a single organism or an isolated part of some larger creature.

(Open University, 1998, p.18)

Example 3

Throughout its 15 billion years, the pace of the universe's development has been accelerating, each new wave of innovation building up to trigger the next, in a series of 'leaps' to further levels of change and diversification. Compress this unimaginable timescale into a single 24-hour day, and the Big Bang is over in less than a ten-billionth of a second. Stable atoms form in about four seconds; but not for several hours, until early dawn, do stars and galaxies form. Our own solar system must wait for early evening, around 6 p.m. Life on Earth begins around 8 p.m., the first vertebrates crawl on to land at about 10.30 at night. Dinosaurs roam from 11.35 p.m. until four minutes before midnight. Our ancestors first walk upright with 10 seconds to go. The Industrial Revolution, together with our modern age, occupies less than the last thousandth of a second. Yet, in this fraction of time, the face of this planet has changed almost as much as at any but the most tumultuous times in the prehistoric past.

(Myers and Kent, 2005, p.12)

Comment

Here are some comments from student readers:

'I haven't studied science before, so some of the words had me puzzled – I'm not sure I could even pronounce Phanerozoic. All the examples talk about the age of the Earth, and how long ago life started. Overall, I think Example 2 is the most difficult, but then, it did say that this was from a higher-level textbook. I had to read that one several times – I hope the rest of the book is easier! I think Example 2 was the most scientific, then Example 1, then Example 3. I underlined some of the words, so I could work out what made sense.'

'Some of the words were new to me, but the confusing part was sorting out all the numbers – billions of this and millions of that. I drew a sort of clock for the last example, so I could see how all the numbers fitted together.'

'I've done some science before, so I wanted to check that the three examples made sense together. They all refer to the age of the Earth as 4.6 billion years, but you need to think about the different ways the numbers are presented. I didn't think Example 3 was very scientific – although it included terms such as "Big Bang".'

'I found the language quite hard to follow – there were lots of long and confusing words within vital bits of information that you needed. I found it took up a lot of time. However the extracts of texts from books were very useful and helpful.'

Your response is likely to differ from all of these, although there may be some similarities. The next information box suggests some ways to deal with unfamiliar language.

Making sense of unfamiliar words and ideas

In science and technology, words have very precise meanings, so you will encounter terms that may be unfamiliar at first. In particular, biology uses a certain style of naming living things. There is more advice and guidance about this later in the book. For now, the first step is to underline or note any unfamiliar language.

Another possibility is to make your own glossary, by writing new words and definitions in a notebook as you work through the chapter. Writing things down in your own words will help your understanding. You will have a handy review of the main ideas that you can refer to later.

In science and technology, text is not the only way to present information. Diagrams, which can be quite complicated, are often used, and it can take time to interpret them. Even so, for some topics, an illustration can clarify the meaning, especially if used alongside the text. The next few figures relate to the three examples in Activity 2. After each figure, I'll describe the content, to give you a guide to interpreting the illustration.

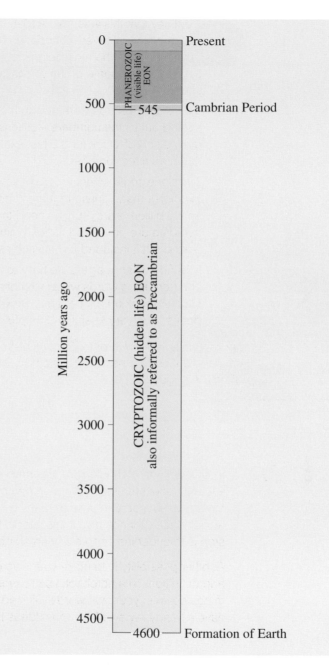

Figure 2 A simplified version of the geological timescale

The geological timescale is divided into Eons, which in turn are divided into Eras and then Periods. To simplify the diagram in Figure 2, most of the technical names for the time intervals have been omitted from this version, which just shows where the Cambrian period fits into the overall scheme. Appendix 1 shows a more detailed version of the geological timescale, which you will need to refer to later in the book. The youngest (latest) events are at the top of the diagram, and the oldest (earliest) at the bottom. The oldest event on this diagram is the formation of the Earth itself. Note that the times are in millions of years. This relates to Example 1 in Activity 2.

Figure 3 Fossils from the early Cambrian period

Figure 3 shows magnified photographs of typical fossils. There is a range of sizes but none of them is longer than a few millimetres. It is not always possible to tell whether each fossil is a complete creature, or just part of one. This relates to Example 2 in Activity 2.

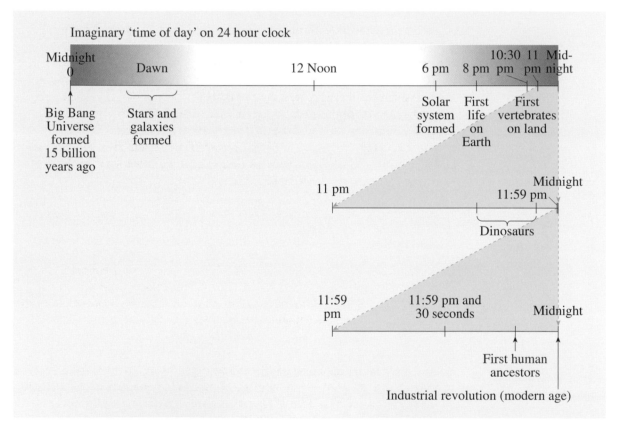

Figure 4 The age of the universe as a timeline

The timeline in Figure 4 shows the oldest events on the left, and the youngest ones on the right. The oldest event in this diagram is the Big Bang, the formation of the universe. This timeline is labelled with the actual time periods in millions of years (below the line), and the imaginary equivalent if the whole sequence were fitted into 24 hours (above the line). The most recent segment of the timeline (the last 'hour') is repeated below at a larger scale, to show more detail, and then the latest segment of that line (the last 'minute') is again enlarged below that. This relates to Example 3 in Activity 2.

Activity 3

Allow about 20 minutes   Using information from diagrams

Read Examples 1, 2 and 3 in Activity 2 again, referring to the illustrations in Figures 2, 3 and 4. How much effect do the illustrations have on your understanding? Would you make more sense of the information if it were only in an illustration, or only in words? Which do you prefer?

Comment

Some people have a preference for words or illustrations, although science and technology use both together. One of the challenges is to interpret the diagrams alongside the text – and there will be opportunities to practise this as you progress through the book. You may also find that sketching your own simple diagrams will help – they do not need to be works of art.

When you have worked through an activity and read the comment, it is worth thinking back on how you got on. Could you have tackled the activity in a different way? How does this fit in with what you have learned from earlier in the course? You might also want to make a few notes on techniques you can use in the future, either for solving problems or for improving your learning. For example, you may decide to try out one of the different ways of taking notes or remind yourself of the importance of using a diagram. An important part of studying is the time spent thinking about what you have done and how effective your methods have been, so that you can improve your work. The next activity helps you to do this.

Activity 4

Allow about 10 minutes   Thinking about your studying

Think about how you felt about Activities 2 and 3 and the comments afterwards. Next time you meet some text with technical jargon, will you approach it differently? What have you learned from tackling the activity about diagrams? Jot down a few notes summarising your thoughts.

Comment

People learn in different ways and it is important for you to try different approaches, to find out which methods work best for you. Some of the ideas in this book are complicated, although we have explained them in some detail, so you may need to spend some more time on some sections.

1.3 What's in this book?

The environment is a vast subject, with many different strands and interconnections. This book covers two main areas: the first half describes some of the connections from a biological point of view, and the second half considers some of the technological implications from a human point of view. The science and technology are woven together, as are local and global examples. We do not claim to be comprehensive, nor to provide any neat answers, but this book should raise awareness, and perhaps form the basis for further debate or study.

There are four themes that run through the book, and are highlighted at various points:

- timescales
- chains and cycles
- local and global
- diversity and sustainability.

Chapter 1 has introduced timescales. Chapters 2 and 3 introduce the biology of life and its interrelationships. This branch of biology, ecology, describes interrelationships between plants, animals and the wider environment. Short case studies and examples, mainly from the UK, illustrate these concepts, including chains and cycles. Chapter 4 discusses the unique role of humans, and some ways in which we have changed the environment to suit our needs, for example, through farming. Historical and contemporary case studies show the impact of human activities on land use, both local and global. Chapter 5 then discusses some of the factors that affect populations of plants, animals and humans. Chapters 6 and 7 move on to a more technological approach, looking at how we can assess the environmental impact of current lifestyles. The ecological footprint (or ecofootprint) is explained in detail with realistic examples. This shows why fuel use is such a crucial issue, in particular fossil fuels such as oil, coal and gas. (The origin of fossil fuels is explained in Chapter 6.) Sustainability is introduced with examples of schemes to reduce environmental impact. Three main areas are considered: transport, waste disposal and domestic energy use. Chapter 8 ends the book, if not the debate, with the question 'what next?' This brings together some of the factors that are involved in the complexity of addressing environmental problems.

Finding information in this book

In common with many other technical books, Introducing Environment has several devices to help you find specific information. The contents list includes subheadings in each chapter, so you should be able to pick out the main topics. The index lists the main occurrences of key words. Where a word is defined in the text, it is highlighted in green in both the text and the index. The study checklist at the end of each chapter lists the study skills and main ideas for that chapter. The reference list at the end of the book gives enough detail for you to trace the sources of the text quotations used in each chapter, if you wish to, although you are not expected to read the original sources. The appendices contain general reference information that should help you at various stages in the book. In addition, the DVD presents information in audio and video format, and there is a DVD icon in the margin whenever you need to use it.

1.4 Science and technology

This book is based on science and technology, which provide a specific perspective on the world. This section introduces these subjects, while later chapters will revisit these topics with other examples.

Activity 5

Allow about 5 minutes

What do science and technology mean to you?

Write a few words or phrases that describe what you understand by 'science' and 'technology'.

Comment

Although 'science' has a more specific meaning, it is not easy to define in a few words. In contrast, 'technology' can mean many different things to different people. Here are some possibilities – you may have thought of others.

Science:

- experiments and theories
- laboratories
- finding out about the world
- testing ideas
- chemistry, biology, physics.

Technology:

- computers and mobile phones
- tools and gadgets
- applied science
- engineering.

In general terms, **science** is about a particular type of knowledge about the world. Scientific knowledge has been tested through experiments and observations. Science involves the investigation, analysis and study of nature. It has its own methods and techniques for finding out things. Before a new observation or theory can be accepted as a scientific fact, it has to be tested by scientists working independently. Some sciences, such as chemistry and physics, tend to be based within laboratories. In these subjects, experiments are often set up under controlled conditions, such as a specific temperature or combination of chemicals. In other sciences, such as astronomy or biology, experiments more often start from observing the world outside the laboratory, and then testing ideas or theories against those observations. The aim is to understand the world around us, to know how things work.

If the aim of science is to comprehend the world, the aim of **technology** is to change it. Technology uses knowledge to achieve a practical purpose, to solve a problem, say, or to satisfy a need. Technology is the application of knowledge, including scientific knowledge, to change things. Often, technology is thought of as applied science, but there are many cases where a technologist has developed a practical solution to a problem before the scientific principles have been fully understood. For instance, people were building boats long before the physical theory of buoyancy (floating) had been worked out. Technology often involves devices or tools, such as power stations or computers, but it also includes social innovations. For example, a book can be thought of as a technology for sharing ideas, or a meeting between people as a technology for sharing experience to solve a problem. Later in this book, farming is used as an example of a technology that involves tools, knowledge and effective organisation of people in order to work.

In practice, science and technology tend to be closely linked. Human beings find it difficult to know something without using that knowledge for some purpose or another. The more we know about the world, the more we are able to change it. But the more we change it, the more we learn about the way things work. Our understanding of our surroundings progresses hand in hand with our ability to change and manipulate them. Understanding the world allows us to change things in new ways, or to change them more effectively and efficiently. Similarly, the process of changing things improves our understanding of how the world works – and thus our ability to investigate it in new ways so as to improve our understanding still further.

An example should make things clearer. People were using yeast to make bread for thousands of years before yeast was identified as a living thing, made of many microscopic cells. The technology of breadmaking had gradually developed as a practical process that worked, without having to know the detailed science. Once the biology of yeast was understood, however, people could apply that knowledge to improve the breadmaking process. So, technology can find solutions before the science is understood, but can also benefit from applied scientific knowledge.

One consequence of this distinction is that we tend to judge science and technology in different ways. Science tends to be evaluated in terms of 'right' or 'wrong', 'yes' or 'no'. Does the Earth rotate around the sun? Yes. Is the Earth supported on the back of a giant turtle? No. In contrast, when we look at technology we ask about whether it works. Is this particular technology effective and appropriate in this particular situation? Does it achieve the desired result?

1.5 Looking back and moving on

This chapter has introduced you to the main learning components of the course and also outlined some study skills. However, the key question is, what have you learned? What do you now know, or can you do, that you didn't know, or couldn't do, at the start of the chapter? When you reach the end of a chapter it is a good idea to review the work you have completed.

Although most people feel fairly confident reading steadily through a chapter from start to finish, there will be situations where it is more important to read through quickly to identify a key bit of information or to look through several sections to get an overall view of the main ideas. In these cases, reading every sentence carefully would not be appropriate and would waste quite a lot of time.

Suppose you wanted to find one bit of information quickly, for example some fact that you knew you had read about in the current chapter. What would you do? One way would be to check in the index – this would give you the page reference and allow you to home in on the information fairly quickly. Alternatively, if the topic was not in the index, you could look at the contents page or flick quickly through the chapters, reading the headings and subheadings. The first sentence in each paragraph often introduces the main idea too. Or if you have a good idea where you read the information initially, you can quickly glance over each page just looking for that one piece of information.

Using a spray diagram to summarise information

To quickly summarise a lot of information, some people find it helpful to draw a 'spray diagram' which shows how the different topics are connected.

Start by writing down the central theme in a 'bubble' in the centre of the page, then for each main idea relating to it, draw a line outwards from the central bubble, and write the idea along that line. Where there are related ideas, draw branches off the main line. In this way you should build up a series of branches showing related ideas. The process of creating the diagram can help to sort out ideas and how they fit together. These diagrams are very personal and can be as elaborate or as simple as you wish – there is no right or wrong way to do it. Some students like to add a lot of detail, for example including colour, pictures, page references and examples, while others prefer a simple plan, concentrating on the key points.

Activity 6

Allow about 15 minutes   Looking back

Summarise what you have studied in this chapter using a spray diagram. Even if you have never used one of these before, have a go, then compare yours with the one in Figure 5 (overleaf). In this case the central theme is 'Chapter 1', so put that in your central bubble then go back through the chapter picking out the main ideas and points that relate to them.

Comment

Don't worry if your diagram looks different from this one. It is your personal record of the content of the chapter and how the different sections relate. You may find connections that others do not spot. You may find it helpful to create your own spray diagram for each chapter and use it to review your work at a later date.

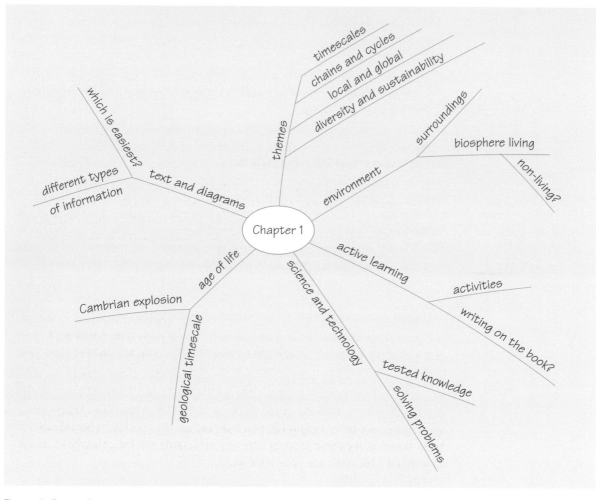

Figure 5 Spray diagram for the contents of Chapter 1

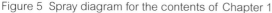

Planning your study time

One of the most difficult aspects of being a student is fitting in your studying with everything else in your life. So it is important both to find enough time to study and then to try to make the most effective use of your time. Finding enough time can be quite a challenge! It often means giving up some activities you currently enjoy or perhaps negotiating with your family and friends to pass on some of the daily chores or to allow you some time to yourself. None the less, it is surprising how much can be achieved in short five- or ten-minute slots, such as recapping on previous work, sorting out paperwork or planning future work. Having found some time, it is also worth thinking about whether this is the best time for you to study and, if it isn't, changing it.

Activity 7

Allow about 10 minutes

Planning your study of Chapter 2

Now quickly read through the main headings in Chapter 2. Using your experience of studying Chapter 1, write a rough plan of the times for your study sessions for Chapter 2. Remember to include some emergency time in case your timetable does not work out as planned or you want to include further work on this chapter.

Comment

A student made the following comment:

> 'I looked through Chapter 2 – there are lots of technical words, like species, that might slow down my reading. Plenty of activities too. I think I'll study this in several chunks. I'll have time on Wednesday afternoon to do a couple of hours – might get as far as the part about black smokers.'

It is worth considering the times of day and the lengths of sessions in which you work most productively. For example, if you know you are going to lose concentration after half an hour or so and also that you are just too tired to study in the evenings, it is probably better to schedule your study time for new topics in half-hour slots in the morning and use the evenings for other chores.

If you tick off the activities you can do, you will be able to see clearly which areas to ask for extra help on or practise further. You will then be able to build sessions for these aspects into your timetable for future study sessions, ensuring a firm basis for your later work.

Study checklist

You should now understand that:

- The term 'environment' has different meanings in different contexts, and this book takes a scientific and technological view of the environment.
- The biosphere encompasses all life on Earth.
- Science is about how the world works.
- Technology is about solving problems.

You should now be able to:

- tackle activities with more confidence
- begin to read actively
- start reviewing your learning and planning further study
- use a spray diagram to summarise information.

2 Ecology and ecosystems

2.1 Introduction

Chapter 1 introduced you to the biosphere. Human beings – you and I – are a unique part of it. Nothing else has our ability to analyse and understand the natural world, or our capacity to change and manipulate it. We alone can ask questions about how the biosphere came to be the way it is. We alone can investigate the processes that maintain it. And, with an ever-increasing sense of urgency, we alone can investigate the effects of our actions and make predictions about the long-term consequences. But with knowledge comes power … and with power comes responsibility.

To study the biosphere, and the importance of our role within it, we must first break it down into smaller, more manageable chunks. These smaller parts – called ecosystems – can then be investigated to identify some of the processes that are common to the biosphere as a whole. That is the task for this chapter.

We will start by looking at ecosystems and at some of the living and non-living components they contain. We will then move on to look at the importance of being able to identify and name the living things in an ecosystem, and examine some of the ways in which we can start to record some of the interactions between them. But we must never forget that we ourselves are members of – and manipulators of – the ecosystems we inhabit. We are a part of the biosphere. We were produced by it; we live within it; and we are beginning to change it in ways we don't yet fully understand. The chapter will end by introducing the concept of 'ecological health', the idea that our growing exploitation of the resources of ecosystems is endangering their ability to support themselves … and us.

But let us start by looking at some of the vocabulary involved. To study or understand anything we have to be clear about what the words we use mean. We will need these words to record what we have discovered, and to share our understanding with other people.

2.2 The science of ecology

This chapter is primarily concerned with a branch of the study of the natural world known as ecology:

> Ecology is the scientific study of the interrelationships between living organisms and the environments in which they live.

Activity 8

Allow about 5 minutes

Words …

Look at the definition of ecology above. As you're already aware, it is customary in science writing to use words in a precise and careful way. So it is important to be able to recognise and understand important terms when you come across them.

Note any terms in the definition that you feel are especially important to its meaning.

Comment

I noted four terms that seemed to be particularly important: 'scientific', 'organisms', 'interrelationships' and 'environments'.

First and foremost, ecology is a **scientific** way of thinking about the world. This means that it involves a certain way of investigating, studying and writing about a topic.

In this context, an **organism** is a living thing – ourselves and other animals, as well as plants, fungi, bacteria and so on.

These living things interact with each other in various ways and with the non-living components that make up the **environment** in which they live. These non-living components include rocks, soils and water, as well as the atmosphere. (The physical locality, the place, in which an organism lives is known as its **habitat**.)

All these interactions produce a complicated set of **interrelationships**. And these interrelationships can take many forms, as we shall see.

Activity 9

Allow about 5 minutes and continue as you read the rest of this chapter

… and meanings

It's not enough to just spot important words, although that's a useful skill. The vital thing is to make sure that you understand what they mean.

Look back at the terms identified in Activity 8. Without looking at the book, can you explain what each one means?

As you work through the remainder of this chapter, you will develop your understanding of the terms 'organism', 'environment', 'habitat' and 'interrelationship'. When you come across one of these terms, stop and think about how it has been used. Does the way the term has been used change, or add anything to, your understanding?

Comment

When you reach the end of the chapter, see the response at the back of the book.

2.3 Ecosystems

There is a problem with any attempt to take an ecological approach to the biosphere as a whole. It is so incredibly complicated and diverse that it is difficult to know where to begin. The number of living and non-living components defies description, and the number of possible interactions boggles the mind. There is also the problem of scale. Do we start with weather patterns that cover the whole globe, examine the impact of human settlements on the wildlife of the Gobi desert, or start by investigating the behaviour of ladybirds feeding on greenfly on a rose bush?

One approach to the investigation of anything big and complicated is to break it down into smaller, more manageable parts. We can then study these smaller chunks at a scale and level of detail that suits our purposes. This approach to complexity is called **reductionism**, because it involves reducing complex things to a collection of simpler parts. We can then take what we learn about how things work in the smaller parts and use it to try to understand the system as a whole.

An **ecosystem** (from 'ecological system') is a collection of living things and the environment in which they live. The size and boundaries of an ecosystem, the bits to be studied, and the interactions to be investigated, are all determined by what we want to know. So an ecosystem can be large (a rainforest) or small (a pond) – and large ecosystems can often be broken down into a number of smaller ones. The important thing is that ecosystems are produced by living organisms interacting with each other and the physical environment.

Remember, all ecosystems involve:

* living organisms
* a physical environment (land, water, air)
* a source of energy to make the whole thing work.

For almost all of the Earth's ecosystems the ultimate source of energy is light from the sun.

Activity 10

Allow about 15 minutes   Constructing a glossary

This section has introduced a number of terms and definitions: for example, ecology, ecosystem, organism, environment, habitat. Some, perhaps all, of these words may be new to you. One way to record them for future use is to construct a glossary, in which you jot down important terms and notes about their meanings.

Use a notebook or a separate page in your study folder. Write down the term and some notes that explain what it means. What you write must mean something to you. Trying to write your own definition, in your own words, will help you to check your understanding. Remember to allow some space so that you can add to, or change, your definitions as you work your way through the course. Think of the glossary as a working document, and don't worry about making it too neat and tidy.

Comment

Here are a few glossary entries produced by one of the student testers who read early drafts of Chapters 1 and 2. They are just examples, and yours may be quite different. The important thing is to think about what the terms mean, and to produce definitions that will remind you of the main points when you return to them.

- Biosphere: the portion of the Earth and its atmosphere that supports life.
- Ecosystem: living things/physical environment/source of energy (e.g. rainforest, pond).
- Reductionism: reducing complex things into a number of smaller parts.

An example ecosystem: the rockpool

Let's examine the concept of an ecosystem in more detail using an example that is familiar to many people: a rockpool on a British beach (see Figure 6).

Like all ecosystems, a rockpool is linked to the wider world and to other ecosystems. This link is most apparent in the shape of two tides every day, which change the sea water in the pool and bring in new organisms from the open ocean (as well as allowing others to escape back into the sea). The tides also change the physical characteristics of the pool and its surroundings. When the tide is out, the rockpool is a collection of organisms living together in a fairly clearly defined place. When the tide is in, the pool may become no more than a small depression on the rocky sea bed.

Now let's look at the components of our rockpool ecosystem in more detail. Remember, any ecosystem contains living things, a physical environment and a source of energy.

The most obvious living things will tend to be the largest ones: seaweeds, sea anemones, whelks, shrimps, fish and so on. But this shouldn't blind us to the importance of the organisms we can't see. For example, the water itself is full of tiny plants and animals – called 'plankton' – that are food for many of the larger creatures. And the water and rocks contain huge numbers of the simple single-celled organisms called bacteria, and other microscopic forms of life, that play an important part in the working of the ecosystem.

The most obvious components of the physical environment of the pool are the rock that surrounds it, the sea water in it, and the air above it (when the tide is out). But these physical factors are far from fixed. The tides, and the effect of sunlight on the exposed pool, mean that the organisms that live in it must be able to withstand changes and extremes of, for example, temperature or salt content. Some small pools dry up altogether in the summer or ice over in winter when the tide is out.

Almost all of the energy that supports the life in the pool arrives in the
form of light from the sun. Some of this energy is captured by seaweeds
attached to the rocks and microscopic plants (types of plankton). These
plants are eaten by animals, which are eaten by other animals, and so on.
But the sun is not the only source of energy involved in the workings of
this particular ecosystem: the tides that sweep across the pool twice
a day are driven up and down the beach by the gravitational pull of the
moon.

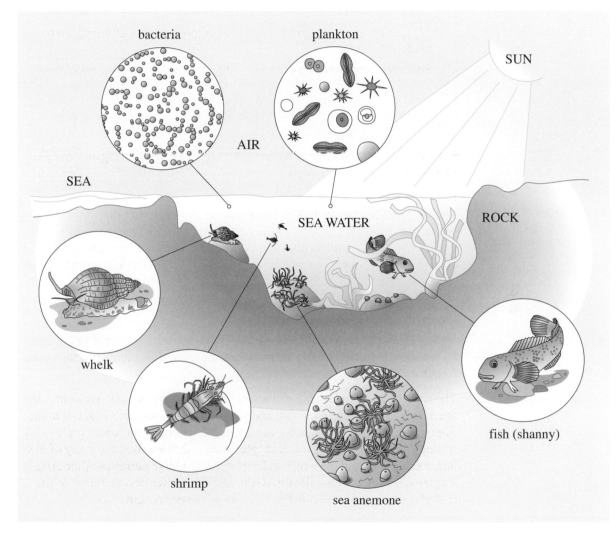

Figure 6 A rockpool

Introducing energy

Energy is a concept that is used in many branches of science. In its everyday meaning, you may recognise it as the 'ability' to do something; whether that is movement, heating, lighting or providing food or fuel. These correspond to different forms of energy – kinetic (motion), thermal (heat), light, and chemical (food or fuel).

Energy can be converted from one form to another. For example, a plant converts sunlight into chemical energy. The energy from the food we eat (chemical energy) is used to keep us warm (heat energy), to enable us to move (kinetic energy) and to enable us to grow or put on weight (chemical energy in muscle and fat).

Although energy can go through many conversions, there is one rule that appears to apply everywhere in the universe. That is, energy cannot be created or destroyed – in any process, the total amount of energy is always constant. You will learn more about energy later in this book.

Activity 11

Allow about 15 minutes

Making notes

This section contains a lot of information. Did you do anything as you read it? Did you simply read the text, or did you make notes?

Read the description of the rockpool again. Try to pick out one or two of the main ideas in each paragraph.

Comment

Here's my attempt.

> Rockpool ecosystem – links to other ecosystems and wider world.
>
> Ecosystem – living organisms, physical environment, energy.
>
> Living things – plants, animals, plankton, bacteria.
>
> Physical environment – rock, sea water, air, tides.
>
> Energy – sunlight (also tides caused by pull of the moon).

Did you note many of the same points? Is there anything else you'd include? Is there anything you'd leave out? How did you set your notes out? Is there room for you to add extra information later on in your studies?

Remember that it is the observer – you or I – who defines an ecosystem, and selects which organisms or aspects of the environment to study. In much the same way, you have to decide what is important when you make notes.

However big or small our ecosystem, and whichever aspect of the system we choose to study, it is ecology that allows us to investigate how it works – and to wonder at the beauty and strangeness of it all.

Black smokers

I started with an ecosystem that is familiar and easy to visualise. Most people have seen a rockpool, if only on television. But many ecosystems remain undiscovered, and some of the ones we know about are full of mystery. The purpose of this section is to introduce one that was discovered fairly recently. Don't worry about the details. It's here as another example of how living things, a physical environment and a source of energy can combine to create an ecosystem.

It is probably true to say that we know less about the bottom of the deep sea than we do about the surface of the moon. However, we know that there are ecosystems down there that demonstrate the interaction between living things and extremes of the physical environment. They are unusual in that the main source of energy is not light from the sun but the heat from molten rock.

At certain points, sea water penetrates cracks in the ocean floor and comes into contact with the lava produced by active volcanoes. The heated water, full of minerals from the rock, rises rapidly to create plumes of hot water called hydrothermal vents ('hydrothermal' is just science-speak for 'hot water'). The minerals in the super-heated water give it a cloudy appearance, which is why the vents are also known as 'black smokers' (Figure 7). The pressures at these depths mean that water temperatures of 300 °C can be found.

Figure 7   A black smoker

Despite the extremes of temperature and pressure there is life here. Bacteria live around the black smokers at temperatures of up to 110 °C, surviving on minerals and the heat energy in the hot water. Other animals feed on the bacteria and each other: blood-red giant tube worms up to three metres long, limpets, barnacles, prawns, crabs and fish. Most of these organisms are found nowhere else on Earth.

Figure 8 A colony of tube worms close to a black smoker

2.4 Making notes

This section develops a useful study skill: making notes. The main scientific story then continues in the next section. Notes can be used for a number of things:

- to develop your own understanding of what you are reading
- to prepare for a specific task, say, a written assignment
- to act as a record for revision.

The first purpose is probably the most important one. After all, it is much easier to write about or revise something that you already understand. Making notes forces you to think about what you are reading and to make sense of it. You have to engage with the text, identify the main points and record them in some way. Good notes pay off in two ways. First, the very process of making and organising your notes improves your understanding of what you have read. Second, your notes will serve as a reminder of the main points long after you have moved on.

The first step is to read the material. Don't worry too much about making notes on the first read through. The important thing is to concentrate on following the line of the argument and making sure that you understand the key points.

Then review the material. Locate the main ideas as well as important points that relate to them. If the book is yours, you can underline or highlight important words or definitions, put a box around important passages, or write notes in the margin.

Alternatively, you can make notes on a separate piece of paper. Remember to leave lots of space so that you can add extra thoughts later on. Learning is a process of changing and developing your understanding. Your notes may take the form of paragraphs, short phrases, bullet points or diagrams (for example, spray diagrams).

The form your notes take will depend on a number of factors:

- the techniques that work best for you
- the nature of what you are reading
- the purpose of your notes
- how much time you have.

Remember that the notes you produce are *your* notes. Experiment to find out what works best for you.

Over time, you will probably develop a range of approaches. So, for example, I tend to use sentences or short paragraphs for material that I find fairly straightforward (see Activity 11). For texts that are more complicated or more difficult to follow I might use a spray diagram. Visual representations often take longer to prepare, but I find the extra investment of time and thought can really help to deepen my understanding. Again, this is a personal preference. Some people find it easier to work with, and to understand, diagrams and pictures. Other people prefer words and lists.

At some point you'll have to give some thought to housekeeping. Where will you keep your notes? In a notebook or a ring binder? I prefer looseleaf notes in a ring binder, as I can add extra pages as necessary, but the choice is up to you. The main thing is to make sure that you can find your notes again when you need them.

Finally, a health warning! Don't put too much time and effort into making notes – and try to avoid writing out the entire text in note form. The trick is to develop an approach that provides the most benefit for the least effort. Always remember that making notes should be a pleasure not a chore. Making notes is an essential part of studying, and if you enjoy the process you'll find you learn a lot more.

2.5 Identification and naming

Before we can start to investigate how the organisms in an ecosystem interact with the environment we have to be able to give them names. The naming of living things has always had an important practical function, allowing us to understand the natural environment and exploit it more effectively. As soon as something has a name, all sorts of useful information can be attached to the label. Is this plant good to eat or poisonous? Is this animal rare or plentiful? The information can then be passed from one member of a group to another, and from one generation to the next.

Scientists face a similar need. To understand an ecosystem we need to be able to name and list the organisms involved in a precise and accurate way. Naming – like reductionism – is a strategy that allows us to impose some order on the complexity of the natural world.

The science writer Richard Fortey explains the importance of this in his book *Life: An Unauthorised Biography*:

> Discrimination and identification have value beyond the obvious separation of edible from poisonous, valuable from worthless, or safe from dangerous. This is a means to gain an appreciation of the richness of the environment and our human place within it. The variety of the world is the product of hundreds of millions of years of evolution, of catastrophes survived, and of ecological expansion. To begin to grasp any of this complexity the first task is to identify and recognise its component parts: for biologists, this means the species of animals and plants, both living and extinct.

> (Fortey, 1998, p.14)

As Richard Fortey notes, the species is the fundamental unit of biological diversity. But what is a 'species'? And how do we distinguish one species from another?

In the field, scientists use two approaches to identify a species. Neither is without its problems.

• Members of the same species normally resemble each other.

• Male and female members of the same species can breed with each other to produce offspring that are also able to reproduce.

The first approach, appearance, is the most obvious one for most of us most of the time. We know that a robin is a robin because it looks like a robin – and the word 'species' is derived from the Latin verb *specere* ('to look at'). But appearances can be deceptive. Males and females of a given species may look different, and many organisms change their appearance as they mature (say, from tadpole to frog or from caterpillar to butterfly).

The second approach is also fairly straightforward in most cases. We take it for granted that our robin can breed with other robins to produce more robins, and that these robins will, in their turn, breed with other robins. But in some circumstances, members of closely related species can and do breed with each other. For example, the horse and the donkey can breed with each other, although the offspring – a mule – is unable to reproduce.

The concept of the species is important to biologists and to our understanding of the working of ecosystems and the biosphere. That's why I've spent some time on it here. The number of species in an ecosystem, or the biosphere as a whole, is an important indicator of its health. Perhaps it is best to think of species as more or less permanent varieties of living things. Many biologists feel that although the definition of a species has its difficulties, in most cases they know one when they see one. The exceptions to the rules are a useful reminder that the complexity of the natural world does not always conform to the categories we attempt to impose on it, and that as a consequence the use of scientific terminology requires judgement and common sense.

Scientific names

Scientists use Latin for the formal names of living things. This means that people from different countries can be sure that they are talking about the same thing. You will come across these scientific names from time to time, so it is useful to know how they work.

Dogs and wolves share a lot of features in common, so they are put together in a genus – a grouping of related species – called *Canis*. They don't normally interbreed, so they are different species: for example, *Canis familiaris* (the domestic dog) and *Canis lupus* (the grey wolf). The Latin names of species are given in italics, or underlined if handwritten. The name of the genus starts with an upper-case letter; the name that indicates the species is given a lower-case letter. All human beings are members of the same species: *Homo sapiens*.

As far as I know, only one animal has a common name that is the same as its scientific name: the boa constrictor (*Boa constrictor*).

2.6 Interrelationships

It is interesting to contemplate an entangled bank, clothed with many plants of many kinds, with birds singing on the bushes, with various insects flitting about, and with worms crawling through the damp earth, and to reflect that these elaborately constructed forms, so different from each other, and dependent on each other in so complex a manner, have all been produced by laws acting around us.

(Darwin, 1859/1985, p.459)

So starts the final paragraph of Charles Darwin's famous book *The Origin of Species*. I have included it here because Darwin sets out beautifully, in one sentence, the complexity of the natural world.

First, he introduces us to an ecosystem – 'an entangled bank'. (I like to think of this as a hedgerow in my home county of Wiltshire. In my mind's eye, I see it on a warm summer's day: a jumbled mix of plants – flowers and trees – with insects crawling through the undergrowth and birds feeding in the branches.)

Darwin notes that this ecosystem consists of different populations of organisms that live together in a particular location or habitat – plants, worms, insects and birds. He also suggests that these organisms are dependent on each other in a number of complicated ways, that is, that they are interrelated. And finally, he writes that these interrelationships are the result of laws – processes – that operate throughout the natural world.

This notion of these interrelationships is so important to the study of the environment that it is worth pausing to examine it in a little more detail.

If a bird in Darwin's hedgerow – say, a robin – eats an earthworm, that's an interaction between the bird and the worm. This interaction has consequences for the bird (a good meal) and rather drastic consequences for the worm (the end of its life). If the worm is to live and play its part in producing another generation of worms, it must avoid being eaten by the robin. But the relationship has implications in both directions. If the bird is to live and help to produce the next generation of robins, it must find and eat a certain number of worms (and other things). This set of interactions produces a link, an interrelationship, between robins and worms.

Of course, the robin and the earthworm will interact with many other living things in many other ways at the same time. The robin will use plant material from the hedgerow to construct its nest, will compete with other robins for territory and mates, and may itself end up as a meal for a sparrowhawk or a cat.

The bird and the worm also depend on, and change, the physical environment of the hedgerow. For example, the bird needs oxygen from the atmosphere and its droppings add chemicals to the soil. The worm extracts nutrients from the soil, and alters its consistency by passing it through its body as it feeds. This makes it easier for the roots of the hedgerow's plants to find the water and the nutrients they need, and so on. It is already easy to see that the interrelationships in our 'entangled bank' are more complicated than we might have thought, and that they involve both the living and the non-living components of the ecosystem.

Do this activity as you
read the rest of this
chapter

Looking for interrelationships

Look out for examples of interrelationships in the remainder of this chapter, especially organisms that rely on others for food. Note any examples in your study folder.

Comment

When you reach the end of this chapter, see the response at the back of the book.

Some interrelationships, like the ones discussed in this section, are relatively straightforward and easy to spot. Others are far harder to see, and the vast majority remain unknown to science. Yet the answers to so many of our questions about the natural world depend on identifying and analysing these interactions. As we shall see, they explain why some animals are common in a given location and others are rare; why some plants are large and others are small; and why some organisms are found in some habitats but not in others. They also explain many of the distinctive characteristics of the physical environments in which organisms live out their lives.

In the next section, we shall look at how scientists go about identifying, studying and recording some of these interrelationships.

Allow about 10 minutes

Reviewing your notes

It is important to go back and review the notes you have made from time to time. Can you improve them in the light of your new understanding and skills? Is there anything that you'd like to add? (This is why it is important to leave some space in your notes.)

Look back at the notes you made in response to Activity 11. Would you make any changes? Will they meet your needs when you return to them later in the book? Think about what your notes are for.

Comment

Making notes is easier if you have a particular goal in mind, for example, an assignment or a question that you want to answer.

2.7 Mapping interrelationships

Now we have started to identify the components of an ecosystem, living and non-living, we can start to look at ways of describing some of the interrelationships between them. As we have seen, one very direct form of interrelationship involves living things eating other living things. Everywhere we look, we see an animal making a meal out of a plant or another animal.

So, having reduced the biosphere to a manageable ecosystem, and named the organisms involved, we can move on to another pillar of the scientific approach to the natural world: observation.

If we return to our hedgerow and look closely at an oak tree, we might observe caterpillars eating the leaves. These caterpillars will, in turn, be eaten by the robin, and the robin may find itself unlucky enough to feature on the menu of a sparrowhawk.

We could describe these feeding interrelationships in words, as I have done in the paragraph above, but this is a rather cumbersome way of recording the information.

We could write it out in notes like this:

> an oak leaf is eaten by a caterpillar is eaten by a robin is eaten by a sparrowhawk

Or we could make it even shorter, like this:

> oak leaf → caterpillar → robin → sparrowhawk

When we use a symbol like → it is very important that we are clear about what it means. In this case the symbol → stands for the words 'is eaten by'. It is important to note the direction of the arrow and to think about what the arrow means.

A diagram like this is called a **food chain**. It is a very useful way of capturing one important aspect of the interrelationships in an ecosystem.

Activity 14

Allow about 10 minutes   Food chains

Rearrange the following organisms from Darwin's hedgerow into food chains.

(a) thrush … grass … cat … snail

(b) sparrowhawk … oak leaf … robin … ladybird … greenfly

(c) rabbit … human being … dandelion

(d) grass … fox … snail … hedgehog

Comment
See the comment at the back of the book.

There are some things to remember about food chains.

First, almost all start with a plant that is eaten by an animal. The chains then vary in length, usually from three to five species. The shortest food chains have just two species: for example, blackberry → human being. We'll look at some of the factors that limit the size of food chains in Chapter 3.

The second thing to note is the presence of human beings, ourselves, in some of the chains. We eat a wide range of plants and animals and occur at the top of a large number of food chains. These chains may involve wild or domesticated plants and animals, but almost all the food in your local supermarket is part of a chain that starts with a plant and ends with you.

The third point to note is that several of our food chains have organisms in common. So, for example, grass and snails are common to lists (a) and (d) in Activity 14. This is not surprising, as most animals eat a number of different plants or other animals. If some of the food chains in an ecosystem are combined they form a **food web**, in which each species appears only once. Food webs make it easier to see how the various organisms depend on each other for food. We can combine all the food chains we have produced for Darwin's hedgerow to create the food web shown in Figure 9.

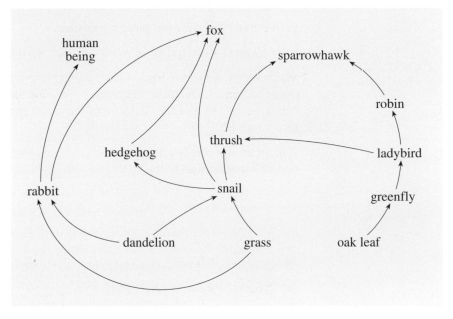

Figure 9 A food web

Allow about 10 minutes   Food webs

Can you add an organism and links to Figure 9 that are not included in our food chains? You may find some ideas on page 28.

Comment

My attempt is shown in Figure 10. You may well have come up with a number of different organisms. Food webs can quickly become very complicated: I have added only one organism.

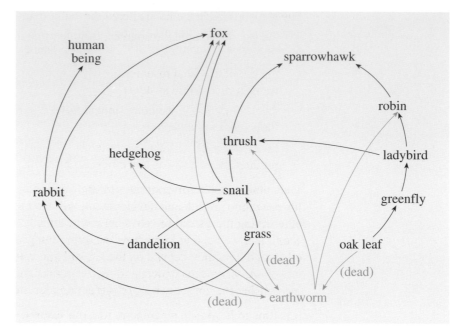

Figure 10   A more complicated food web

2.8 Ecological health

Does all this matter? Why is it important to divide the biosphere into ecosystems and study the interrelationships they contain?

My first answer would have something to do with the wonder and fascination of nature itself. For me, and for many other people I suspect, what science can tell us about how the biosphere works – about the interrelationships involved – serves to make it more interesting and more beautiful.

There is, however, another pressing reason to study ecosystems. Preserving the ecosystems of the Earth, and their ability to sustain us, is now our responsibility. It is time to put ourselves back into the picture. If asked, we define ourselves in terms of nationality or employment or social class. But we belong to ecosystems as well, and the adaptability of human beings means that we can be found in all sorts of environments from the poles to the equator.

As we have seen, all living things shape – and are shaped by – the physical environments that they inhabit. But no other species has our capacity to alter the world around it, to maximise the exploitation of its resources. And our influence is now global: modern patterns of business, food distribution and the use of natural resources mean that most of us have an indirect influence on ecosystems across the world.

The ways in which we have altered the ecosystems of the world include:

* the use of biological resources, from hunting and fishing to cutting down forests
* the use of physical resources, such as quarrying for rock and diverting water for irrigation systems
* the use of energy resources, including the burning of wood and fossil fuels (coal, oil and gas)
* the creation of artificial ecosystems, such as agricultural land for food production, and towns and cities as places for us to live.

The human population continues to grow and this, combined with the pressure for economic growth and development, will tend to increase our demands on other living things and the physical environment. To manage – or protect – an ecosystem we need to know how the living things it contains depend on each other, and how they depend on the air, soil and water in which they live. In large part, this means understanding the interrelationships involved, and so recognising the consequences of our actions for the ecosystem as a whole.

It is time to introduce a new term into our examination of ecosystems: **ecological health**. This term has become increasingly popular in discussions about the environment, although it is difficult to define or measure with any accuracy. It is possible to identify specific changes to an ecosystem, but any evaluation of its health is a matter of judgement not fact. Who decides whether an ecosystem is healthy? And on what grounds? Is a desert as healthy as a rainforest? Or does ecological health simply depend on a lack of human interference?

The fact that we find it hard to define ecological health doesn't mean that it has no value as a concept. We find it hard to define human health, but we recognise that it is important none the less. We know that some individuals are healthier than others and that certain things – a poor diet, for example – can have a negative effect. Like human health, ecological health is best thought of as a combination of many different things: the diversity, numbers and condition of the living organisms in an ecosystem; the complexity of the food webs involved; and the quality of the air, soil and water that make up the physical environment.

We need to study the health of ecosystems to find out how to protect them. How much change has already taken place? What will be the long-term consequences of our actions? How can we increase an ecosystem's ability both to resist change (**ecological resistance**) and to recover from the changes that have already happened (**ecological resilience**)?

We are the most powerful actors in most ecosystems, yet until recently we have been largely unaware of the ecological consequences of the way we live our lives. A quick glance through the newspapers, however, indicates that we are now becoming increasingly concerned about our collective impact on ecological health, in terms of pollution, climate change and the use of finite biological and physical resources.

Don't forget to compare your notes for Activities 9 and 12 with the comments at the back of the book.

Study checklist

You should now understand that:

- Ecology is a scientific approach to the study of the biosphere.

- Ecosystems are created by the interrelationships between living organisms and the physical environments they inhabit (land, water, air). Ecosystems require a source of energy to make them work and for most, although not all, this is light from the sun.

- To study ecosystems we have to start to identify the components involved and the interrelationships between them. We can list the living organisms by identifying the species involved.

- Food chains and food webs are a way of mapping one type of interrelationship between the organisms in an ecosystem.

- Human beings are part of ecosystems, as well as manipulators of ecosystems. As such we are dependent on, as well as responsible for, the ecological health of the ecosystems we inhabit.

You should now be able to:

- construct a glossary of scientific terms

- make notes on what you read and review your notes.

3 Energy flows and agriculture

3.1 Introduction

Ecosystems need energy to work, and all living organisms need energy to survive. In this chapter we will revisit the food chains from Chapter 2 to investigate how living things process that energy in the form of food, and what this can tell us about the interrelationships between the living and non-living components. We shall also examine the implications for our own use of ecosystems to produce the food we need to support a growing human population.

Energy enters an ecosystem from outside, usually in the form of sunlight, and is eventually lost from it in one way or another. The chemicals necessary for life – for example, oxygen, carbon dioxide and water – are available in fixed amounts in the physical environment and have to be used over and over again. (Carbon dioxide is sometimes referred to by the abbreviation 'CO_2'. In this book, I will write out 'carbon dioxide' in words, as chemical notation is not covered here.)

Activity 16

Allow about 15 minutes

Turning your notes into sentences

One reason for making notes is to prepare for writing an assignment or an essay. The important thing here is that you understand what you have included in your notes, and can transform them back into written text in response to a specific question.

To explore this, take a short section of your notes and try to write it out in the form of a few sentences. Try to make each sentence convey a single point, or perhaps two related points. Avoid sentences that are too long or complicated.

Comment

After you've written your sentences – but not before – look back at the relevant section of this book and compare what you have written with what is in the book. Is what you have written clear? Does it convey the main points?

Did you find it easy to select the main points from your notes? If not, how could you change the way you make your notes to make it easier in future?

3.2 Producers: making your own food

Plants are the essential foundation for almost all life on Earth, and they dominate the land and the sea.

The plants of the land range from tiny mosses to large trees. At first sight the oceans seem to contain little plant life apart from the seaweeds that grow in shallow water where the sea meets the shore. But, again, appearances can be deceptive. The oceans are full of microscopic plants that float in the surface layers of the water. These plants are green, like most of their relatives on land, but they are too small to change the colour of the water.

(a)

(b)

(c)

(d)

Figure 11   The variety of plant life: (a) cactus (b) trees (c) sunflowers (d) microscopic plankton

Activity 17

Allow about 5 minutes

Reviewing your glossary

This chapter will also introduce a number of terms and definitions.

Look back at the glossary you started in Chapter 2. Is the meaning of each term still clear? Is there any information you'd like to add?

Remember to record any new or unfamiliar terms, and some notes about their meaning, in your glossary as you work your way through this chapter.

Photosynthesis

Plants have an advantage over the animals, a trick of their own. They can 'produce' their own food.

Plants construct their food out of simple building blocks in a process called **photosynthesis**. They take carbon dioxide from the air and water from the soil and combine them to make sugars, releasing oxygen into the air as a by-product. The whole process is driven by energy in the form of sunlight, which is why plants need light to grow. (This explains why the plants of the open ocean are so small. Only tiny plants are light enough to float in the sunlit upper waters where they can get the light they need to photosynthesise.)

Activity 18

Allow about 10 minutes

A word diagram for photosynthesis

We can use diagrams to summarise the process of photosynthesis in much the same way that we used diagrams to represent food chains in Chapter 2. Fill in the blanks in the following outline to produce a word diagram that shows how plants combine raw materials to produce the products of photosynthesis.

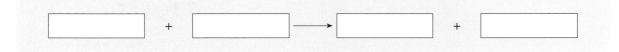

Words: sugars, water, carbon dioxide, oxygen.

What do you think the arrow means in this diagram?

Comment

The word diagram for photosynthesis is:

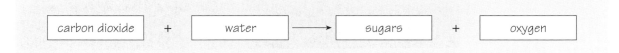

This process *requires* energy in the form of sunlight.

The arrow in this diagram means something like 'result in', 'produce' or 'are transformed into'.

Plants convert some of the energy in sunlight into sugars using a chemical called **chlorophyll** (pronounced 'clora-fill'), a green pigment found mainly in the leaves. It is chlorophyll that makes the world of plants a green one, and explains why we describe a good gardener as someone with 'green fingers' or someone who campaigns on environmental issues as a 'green' activist.

Chlorophyll is usually concentrated in leaves that act like solar panels and are shaped and arranged to catch as much precious sunlight as possible. Some plants, trees for example, put a lot of effort into growing taller than their neighbours to gain more light. But the tree has to build other structures, such as a woody trunk, to support the canopy of leaves. Other plants, such as grasses, use a different tactic. These plants try to cover as much of the surface of the ground as they can, expanding horizontally rather than vertically.

Autumn colour

Some of the trees in Darwin's hedgerow will be deciduous – that is, they will lose their leaves in the autumn and produce new ones in the spring. These plants give up photosynthesis in the cold and darkness of winter and survive on the food they have manufactured during the long, light days of summer. Deciduous trees take the precious green chlorophyll out of their leaves before they fall in the autumn. The brown, red or yellow chemicals that remain in the leaves are what give these trees their autumn colour. (As their name suggests, trees that are evergreen keep their leaves, their chlorophyll and their colour all year round.)

The substances that remain in the leaves of deciduous trees to give them their autumn colour include chemicals called 'carotenoids', which are yellow, orange, brown or red. By a strange coincidence, carotenoids are also important for the colour vision in our own eyes. So there is a direct link between the colour of our hedgerow in autumn and our ability to see, and appreciate, it.

Organisms such as plants that can make their own food are called **producers**, because they can 'produce' new living material from non-living raw materials in the air, soil and water. (We have already come across another producer. Remember the heat-loving bacteria that live next to the black smokers in Chapter 2? They are also producers, using the chemicals and energy of the hot water to manufacture their food.)

Respiration

Some of the sugars produced by photosynthesis are converted into proteins to build tissues that the plant needs for growth and reproduction. Some are converted into other carbohydrates, such as starch, that the plant can store for future use. But a lot of the sugars produced are used to provide the plant itself with energy. Having produced its food by photosynthesis, the plant breaks it down again to release the energy it needs to fuel the processes that keep it alive and growing. The process of breaking down foods to release the stored energy they contain is called **respiration**.

All living things respire all the time to keep themselves alive. Respiration requires oxygen and produces carbon dioxide and water as by-products.

Activity 19

Allow about 10 minutes   A word diagram for respiration

Produce a word diagram for respiration, similar to the one for photosynthesis in Activity 18, using the same words.

[_____]   +   [_____] ⟶ [_____]   +   [_____]

Words: sugars, water, carbon dioxide, oxygen.

Compare this word diagram with the one for photosynthesis. What do you notice?

Comment

The word diagram for respiration is:

| sugars | + | oxygen | ⟶ | carbon dioxide | + | water |

This process *releases* the energy stored in the sugars.

The word diagrams for photosynthesis and respiration mirror each other.

We can summarise photosynthesis and respiration using the following word diagram:

| carbon dioxide | + | water | ⟷ | sugars | + | oxygen |

In photosynthesis, light energy absorbed by the plant drives the process from left to right: making sugars *requires* energy.

In respiration, the process goes from right to left: breaking down sugars *releases* energy.

Plants produce more oxygen through photosynthesis than they use in respiration. This is just as well, as animals – including human beings – depend on this extra oxygen for their own respiration. All animals depend on producers, usually plants, for the food they eat and the oxygen they need to break down that food to release some of the energy it contains. Putting it simply, almost all life depends on plants.

3.3   Consumers: using other organisms as food

Unlike plants, animals cannot make their own food. They have to obtain the energy and the raw materials they need by eating plants or other animals and releasing some of the energy stored in what they eat through respiration. Organisms that use other living things as food are called consumers.

In any ecosystem, the organisms that can make their own food – producers – are almost always plants. This is why plants represent the foundation of most ecosystems. Some of the energy trapped by photosynthesis and stored in the plants is transferred to the animals that eat them. So the primary consumers are herbivores, animals that eat plants. The **secondary consumers** and **tertiary (third-level) consumers** will be carnivores, animals that eat other animals. Food chains also contain omnivores, animals that eat both plants and animals, as we human beings do.

Detritivores and decomposers

There are two particular types of consumer that are important to any ecosystem. A **detritivore** is a scavenger that eats dead organisms (for example, a vulture or a carrion beetle). A **decomposer** breaks down dead organisms outside its own body and then absorbs the resulting nutrients, before breaking them down even further to release the stored energy. The decomposers are mainly bacteria and fungi, and they are responsible for the rotting of dead plants and animals.

(a) (b) (c) (d)

Figure 12   Detritivores and decomposers: (a) vulture (b) dung beetle (c) mould on tomato (d) bracket fungus

Fungi – such as mushrooms, mildews and moulds – are not plants. They don't possess chlorophyll so they are unable to make their own food. Some bacteria can photosynthesise and make their own food, but many species, like fungi, act as decomposers.

The role of detritivores and decomposers is vital as they complete the cycle of growth, death and decomposition. They break down the bodies of dead plants and animals to release simple nutrients containing carbon, nitrogen and other chemical elements. These nutrients then go back into the soil, air and water to be used in future plant growth. Without these nutrient cycles, ecosystems would quickly run out of the basic chemicals necessary for life. If there were no carbon dioxide in the atmosphere, for example, there would be no photosynthesis, no plants and no food for the animals that eat the plants.

Bacteria are especially important in the recycling of nitrogen. Plants depend on nitrogen to make proteins, which are essential for health and growth. Unfortunately, however, plants are unable to use the nitrogen in the atmosphere. But some bacteria can. The main way in which nitrogen becomes available to plants is through nitrogen being 'fixed' by these bacteria, that is, incorporated into a range of nitrogen compounds. Some of these nitrogen-fixing bacteria live free in the soil, but others live within the tissues of certain plants (for example, on the roots of beans, peas and clover). When these bacteria and the plants that they live in die and decompose, the nitrogen compounds leak out into the soil and can be taken up through the roots of other plants.

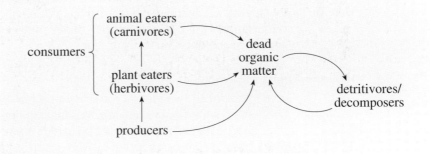

Figure 13   Some of the interrelationships between producers, consumers, detritivores and decomposers

Bacteria

Bacteria exist in huge numbers in almost all environments and their activities are vital to the working of ecosystems. They were among the first living things to evolve on the Earth, some 3.5 billion years ago, and they remain the most abundant form of life. The amount of life that exists in the form of bacteria is far greater than all the other living things combined, whether measured in number of species, number of individuals or sheer combined weight. We tend to concentrate on the larger organisms – the bigger plants and animals – because they exist at a scale we can see and appreciate. But life on Earth remains, as it always has been, predominantly bacterial.

A teaspoon of soil contains over a billion bacteria, and different species can be found from the tops of the highest mountains to the bottom of the deepest seas. As we saw in Chapter 2, heat-loving bacteria can live in water at temperatures of over 100 °C, while other species live in the frozen rocks and ice of Antarctica. One species – scientific name *Deinococcus radiodurans* –can even survive a radiation blast a thousand times greater than the lethal dose for a human being.

We ourselves provide a home – an ecosystem – for hundreds of different species of bacteria. The number of bacteria we carry around with us is far greater than the number of cells in our bodies, and there are thousands of bacteria on a square centimetre of the skin of your hand. The vast majority are harmless, and many are positively beneficial. In fact, we could not exist without them. Bacteria in our intestines help us to digest our food, and some of the 'friendly' bacteria we provide a home for protect us against other, harmful, bacteria that can cause disease.

Activity 20

Allow about 20 minutes

Compost heaps

Garden gold

Listen now to Track 1 on the DVD, which reveals just how much scientific information can be obtained from the workings of a garden compost heap. The detritivores and decomposers in a compost heap break down dead plant material to release nutrients that can be added to the garden, in the form of compost, to help the growth of new plants. These plants may, in turn, find themselves on the compost heap in the future. In this way a compost heap demonstrates the cycles of growth, death and decomposition that recycle important nutrients in all ecosystems.

As with reading this book, the important thing is to think about the process of learning from the material on the DVD. What will you do to ensure that your mind doesn't wander as you listen, and how will you check that you understand what you are hearing? Will making notes help, and what notes or diagrams could you make that will be useful to you in the future?

You may find it helps to listen to the track straight through to get some idea of the main points. You can then think about how you will approach making notes, before listening to the track again with a pen in your hand.

As with making notes from texts, there is no one right way to approach this task. The important thing is to find an approach that works for you.

Comment

You may not have given compost heaps much thought before, but I hope the interview on the DVD has prompted you to take a 'fresh look'. Think about how some of what has already been discussed applies to this new topic – interrelationships and the activities of decomposers, for example.

A new thought may have occurred to you, one that I'll come back to later in this chapter. Ecosystems, such as compost heaps, can be altered, manipulated or managed for a particular purpose.

Bone-eating zombie worms

As with black smokers, an example that is less familiar than the humble compost heap may help to make the role of detritivores and decomposers in ecosystems a little clearer. Let's go back to the bottom of the sea.

For a long time it was assumed that the mudflats that form the floor of parts of the deep ocean must be lifeless deserts. Away from the black smokers the water is too cold and too dark for photosynthesis, so there is little in the way of producers to form the starting point of conventional food chains. Without producers there can be no consumers, although a few organisms were known to eke out a living on a fine mist of dead and living microscopic organisms that drifts down from the warmer, lighter waters above.

Every so often, however, the situation is changed for a small region of the sea floor by the arrival of a more substantial gift from above. The body of a dead whale, perhaps killed by starvation or exhaustion on one of its long migrations, sinks down through the water and settles gently on the mud. The arrival of this huge package of energy-rich material into such a nutrient-poor environment is enough to attract a rich and varied community of opportunistic consumers: detritivores and decomposers.

The first creatures to find the body are usually general scavengers – detritivores – that travel the ocean looking for finds like this one. These 'vultures' of the deep sea, including lobsters and hagfish, home in on the corpse and start to strip off the soft parts. Even so, it can take up to 10 years to remove all the meat from the bones of a large whale.

When the general scavengers have finished the specialists move in. These animals feed on the bones and blubber of the whale. The skeleton quickly becomes covered with a writhing mass of hundreds of different species of animal, including worms, snails, clams and limpets. Mats of bacteria form on the mud surrounding the body and hoover up the nutrients that leak out into the water.

Some of the organisms scientists have discovered appear to specialise in feeding on whale carcasses and are found nowhere else. For example, bone-eating 'zombie worms' – a few centimetres long – end in long root-like structures that can penetrate into the bones themselves to suck out nutrients. The worm's 'roots' contain bacteria – decomposers – that dissolve the bone structure, allowing the worm to digest the results. When the worms have finished, the last useful chemicals are extracted by other, free-living bacteria that invade the interior of the bones.

It is thought that the whole process can take up to 100 years.

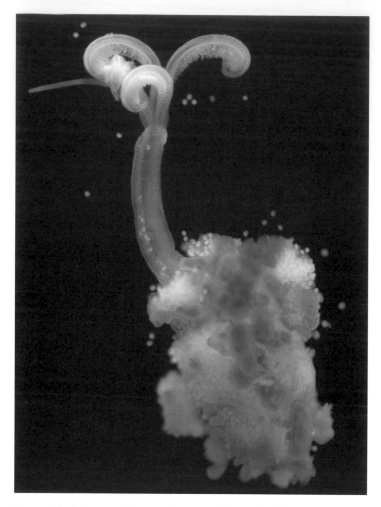

Figure 14   A bone-eating zombie worm. (The scientific name of one particular species is *Osedax mucofloris*, which translates as 'bone-eating snot-flower')

3.4 Food chains revisited

Now we've learned about producers and consumers we can revisit the food chains and food webs from Chapter 2 and add some useful information to our interrelationships.

The feeding relationships in our food chains represent the way in which nutrients and energy are transferred from one organism to another. The stages in the food chain – from producer to consumer and so on – are called **trophic levels**. The arrows in Figure 15 now also represent the flow of nutrients and energy along the food chain.

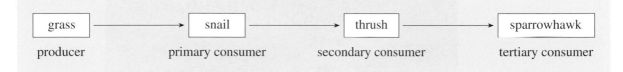

Figure 15   Food chain showing trophic levels

If we go back to Darwin's hedgerow and think about the numbers of organisms at each trophic level we notice something revealing.

The first things we see are the producers – plants. Plants are everywhere, from grasses and small flowering plants to the trees that grow at intervals along the hedgerow. If we look closely, we may start to spot the large numbers of snails, insects and other animals that feed on the plants – the primary consumers. From time to time we'll see birds such as thrushes that feed on the snails and insects – the secondary consumers. If we're very lucky, and spend a long time looking, we might see a single sparrowhawk – a tertiary consumer – hunting for some of the smaller birds.

At each trophic level after the producers, the number of organisms tends to go down: lots of insects and other plant feeders, a few snail- and insect-eating birds and the occasional sparrowhawk. The average size of the organisms tends to increase: thrushes are bigger than snails, and sparrowhawks are bigger than thrushes. There are exceptions to this number and size observation, but it seems to hold true for a lot of food chains.

Why is this? Why aren't there lots of sparrowhawks?

To start to answer this question – and to start to understand our own manipulation of ecosystems as sources of food – we have to think about the amount of energy available at each trophic level. We have to move from observation to analysis.

3.5 Food pyramids: numbers, biomass and energy

To understand more about energy in our feeding relationships, we need to know how much food is available at each step in a food chain.

We could start by looking at the numbers at each trophic level in Figure 15. We have our grass plants and, above them, a **numbers pyramid** showing the drop in the number of organisms at each trophic level (Figure 16 overleaf).

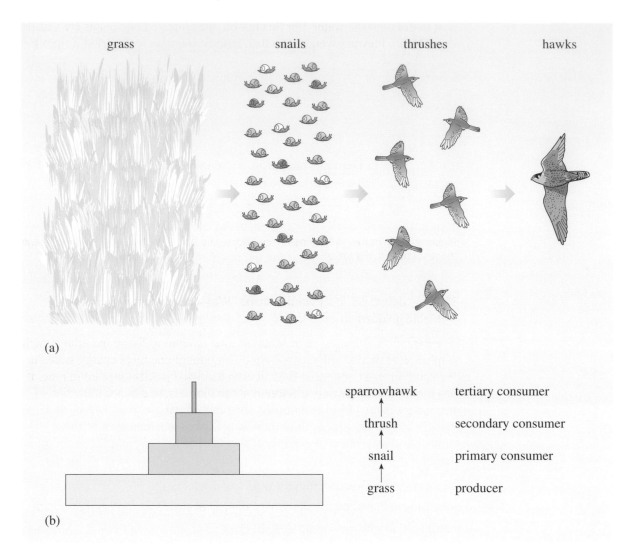

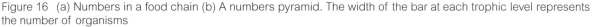

Figure 16 (a) Numbers in a food chain (b) A numbers pyramid. The width of the bar at each trophic level represents the number of organisms

In some circumstances, however, looking at numbers can be misleading. For example, a single oak tree can support many caterpillars. If we want to know how much food is available at each level, we have to analyse the amount of living material involved.

One way of showing this information is to measure the weight – or biomass – of the organisms at each trophic level. This is the living material that is potentially available as food to the consumers at the next level. Biomass is measured in grams or kilograms, like any other mass.

So if we were to collect together the organisms at each trophic level and measure the mass we would get a biomass pyramid like the one shown in Figure 17. (Living things contain a lot of water, which they can't use as a source of energy. In discussing the flow of energy through a food chain, it is best to discount the water. For this reason, the collected organisms are usually dried before they are weighed so that, strictly, biomass is the dried weight.)

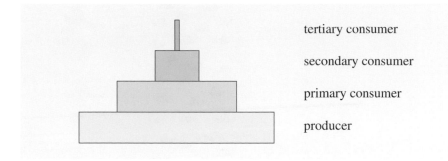

tertiary consumer

secondary consumer

primary consumer

producer

Figure 17 A pyramid of biomass. The width of the bar at each trophic level represents the amount of living material

But why do we get this pyramid shape? Why does the amount of food available go down at each level?

The pyramid of biomass represents the total amount of living material at each trophic level. It also represents the total amount of chemical energy stored in the plant material or animal flesh at each level at a particular point in time. It is a pyramid because only a fraction of the biomass, and hence a fraction of the energy, passes from one trophic level to the next.

Activity 21

Allow about 10 minutes   Energy and trophic levels

Why do you think that the amount of energy available as food is reduced at each trophic level?

Comment

There are a number of reasons.

- Only a proportion of the food at each level is gathered by the next level up: plant eaters never eat all the plants, and meat eaters never catch all their prey.

- Consumers rarely eat all of the organisms they feed on, or digest all of what they eat.

- Living things use energy continuously to stay alive, and lose energy in the form of heat and waste products.

So our pyramid of biomass can be used to produce a pyramid showing the amount of energy passing from one trophic level to the next (Figure 18 overleaf).

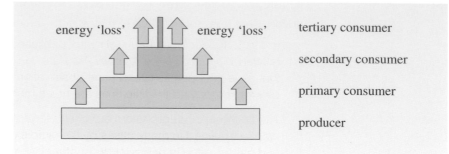

Figure 18 Energy pyramid, showing energy lost from the food chain. The width of the bar at each trophic level represents the amount of energy at that level

There is a large loss of energy at each stage along the food chain. At each trophic level, only about one-tenth of the energy received is converted into the chemical energy in biomass available as food for the next level up. The energy lost from the food chain is not destroyed – according to the law of conservation of energy, energy cannot be created or destroyed (remember Section 2.3). The bullet points in the comment for Activity 21 show what happens to the other nine-tenths of the energy.

This explains why food chains rarely have more than four trophic levels. If we start with 1000 units of energy at the level of the producers, only one unit of energy will be available to a tertiary consumer (the sparrowhawk in our example).

Figure 19 Energy transfer in a food chain

Energy pyramids allow us to look at ecosystems in a new and revealing way. By looking in detail at just one aspect of the interrelationships in our hedgerow – food chains – we can now see that the amount of energy available goes down from one trophic level to the next. Diagrams often work like this. They take something complicated and represent it in a way that is easy to understand. In part, they do it by pulling out just the information required to make the point from a more complex – and complete – picture. (Spray diagrams can be used to make sense of complicated texts for much the same reason.)

By highlighting energy flow, energy pyramids show us that the number of plant eaters (primary consumers) in an ecosystem will be determined in part by how many plants (producers) there are for them to feed upon. The number of secondary consumers will be determined in part by the number of primary consumers, and so on up the pyramid.

Why big, fierce animals are rare

In 1978, the ecologist Paul Colinvaux published a collection of essays entitled *Why Big, Fierce Animals Are Rare*. He started the essay that gave the book its title with the observation that 'Animals come in different sizes, and the little ones are much more common than the big.'

The first point, that predators are often bigger than the animals they eat, is explained in many cases by the simple necessities of catching and eating, often in one swallow, another living thing. The second point, that large predators are rare, is explained by the energy pyramid. The organisms at each level have to survive on the food (energy) that they can obtain from the level below. In Colinvaux's words:

> For flesh eaters, the largest possible supply of food calories that they can obtain is a fraction of the bodies of their plant-eating prey, and they must use this fraction both to make bodies and as a fuel supply. Moreover their bodies must be big active bodies that let them hunt for a living. If one is higher still on the food chain, an eater of a flesh-eater's flesh, one has yet a smaller fraction to support even bigger and fiercer bodies. Which is why large fierce animals are so astonishingly (or pleasingly) rare.
>
> (Colinvaux, 1978, pp.22–23)

The energy pyramid has the surprising consequence that the numbers of prey play a role in determining the numbers of predators. The survival of the organisms at each trophic level is dependent on the amount of food available from the trophic layer below it in the energy pyramid.

Activity 22

Allow about 5 minutes

Time management

Now might be a good time to start to think about how you are managing your time. We all lead busy lives, and studying can appear to be just one more thing to fit into an already overcrowded day. Your success as a student will depend, in part, on how you juggle the demands on you to ensure that you have the time – and energy – you need to study. Think about how you have studied this book so far. Did you read continuously for a long time, or did you study a few pages at a time?

Comment

There are a number of time management techniques, and some experimentation may be necessary to find out what works best for you. Your situation will be unique, so there is no approach that will suit everyone. The following principles may help.

Note tasks in a planner and include the date by which they have to be completed. Plan the week ahead if possible. Do this at the same time each week.

Plan enough time to complete tasks. Be realistic, and be flexible enough to allow for unexpected events. Try not to plan really long study sessions; they may be counter-productive if your attention begins to wander.

Set clear aims. Ask yourself what you want to complete/research/revise.

Prioritise what needs doing: what must be done and what could be done.

Shorter, regular sessions are far more likely to be effective than one long one.

Make use of information technology (IT) if you have access to it and can use it efficiently. For example, word processing a written account allows you to draft and redraft it without having to write it out again each time. You could also use a word-processing package for your glossary, adding to or revising your definitions as you work your way through the course.

Review what you've accomplished at the end of each study session. Highlight anything that you were unable to finish.

Of course, we human beings are members of ecosystems … and the energy pyramid has implications for our interactions with the natural world.

3.6 Agriculture: altered ecosystems

As in Chapter 2, it is now time to look at ourselves as members of ecosystems. We change and direct the properties of ecosystems for our own ends, and nowhere is this clearer than in our use of animals and plants for food.

As with any consumer, our ability to feed ourselves is dependent on the food (energy) we can obtain from the trophic levels below us. The difference is that we can change the properties of ecosystems – if not the rules – to maximise our exploitation of the energy pyramid. The numbers are unclear and controversial, but of the 6 billion people currently living in the biosphere, it appears that the lives of approximately 1 billion are cut short by not having enough to eat … and the lives of another billion or so are shortened by having too much to eat (or too much of the wrong things).

In ecological terms, the earliest humans were much like any other species. They were part of a number of food chains and energy pyramids, consuming plants and other animals and being consumed in turn by a number of predators. However, things began to change. The development of language allowed us to collaborate with each other. This, along with the use of tools and fire, allowed us to collect and prepare a wider range of foods.

Agriculture, by which I mean the cultivation of plants for food by human beings, was a critical stage in the development of human civilisation. Humans have always depended on plants and animals for food, but for most of our evolutionary history we have been hunter-gatherers, collecting wild plants and hunting as we moved from place to place. Hunter-gatherers are nomadic, as they quickly deplete the potential of local energy pyramids and have to move on to new territory. Agriculture allowed us to harness the food potential of plants and animals more efficiently. In effect, it allowed us to broaden the base of the energy pyramid so that it would support more of us ... but this has had consequences for our way of life and for our interactions with the ecosystems involved.

Agriculture involves creating a managed ecosystem in which some of the factors we have been exploring in this chapter – trophic levels and energy pyramids – are manipulated to produce lots of energy-rich foods for human consumption. To do this, farmers have bred species of plants and animals that grow quickly and are resistant to disease; they also control the conditions under which the plants and animals live in order to maximise food production. Successful agriculture depends on converting as much of the potential biomass of plants or animals as possible into appropriate foodstuffs, ideally without reducing the long-term ability of the ecosystem to produce food in the future. Farmers, unlike hunter-gatherers, tend to stay put.

Activity 23

Allow about 10 minutes

Food production

What factors will affect the food production of a plant crop?

Comment

You will probably have come up with some of the items in the following list:

- the amount of energy received by the plants in the form of sunlight
- the supply of water
- the supply of atmospheric carbon dioxide
- the availability of nutrients in the soil
- the amount of the plant that can be harvested and used as food
- the amount of the crop lost to pests or disease
- temperature will also be a factor for some plants.

Arable agriculture, growing plants for food, can be seen as a process in which ecosystems are manipulated to take account of all these factors. The idea is to make sure that as much of the biomass of the crop as possible passes up to the next trophic level – human beings.

Seeds are sown with enough space for the plants to grow, and trees and hedgerows may be removed to ensure that the crop receives as much light as possible. Extra water is made available, if required, through irrigation systems, and in some intensive agricultural practices carbon dioxide is pumped into greenhouses. The fact that so little of the crop is available to decomposers means that chemical nutrients – artificial fertilisers – have to be added to the soil to replace the nutrients removed every time a crop is harvested. In some cases, selective breeding has been used to ensure that a large proportion of the plant is devoted to the seeds, fruit, leaves or roots that we use as food.

Herbicides (chemicals that kill plants) are used to ensure that weeds do not compete with the crop for nutrients, light or water; and pesticides (chemicals that kill animals) are used to control other consumers that would eat the crop before it is harvested for our use. Machinery is used to make the whole process, including harvesting, more efficient – and large areas may be devoted to a single crop or monoculture for much the same reason.

Similar methods are used to maximise the food – meat – produced by the animals we farm. In intensive meat production, animals may be kept in heated sheds and their movement may be restricted to prevent them from using up too much energy keeping themselves warm or chasing each other around a field. Dietary supplements and hormones may be used to increase the accumulation of body mass, and antibiotics may be administered to prevent disease.

Agriculture and ecological health

There are those who feel that intensive agriculture may be unsustainable in the long term, because of its implications for the ecological health of the ecosystems involved. Chemical fertilisers can damage the soil and prevent the natural recycling of nutrients. Pesticides kill harmless organisms as well as harmful ones, and lead to the deaths of other animals in the food chain by killing the animals and plants they feed on.

One approach, labelled organic farming, tries to reconcile our growing demand for food with a concern for the health of ecosystems. Organic farmers recycle nutrients naturally using manure and crop rotation (described in Chapter 4), and try to control pests using natural means (including natural predators).

Another recent development has been the use of genetic modification (GM) to create plants and animals that are more productive. Altering the genetic make-up of organisms in this way remains controversial, and the advantages and disadvantages remain unclear. Over 80 per cent of the maize grown in the USA has been genetically modified in some way, to improve yields or introduce a resistance to pests of various sorts.

Activity 24

Reviewing your time planning

Look back at the time planning you established in Chapter 1. Is it working? Have you experienced any problems? If so, how did you deal with them?

Comment

Here is what one of our student testers wrote:

> 'In Chapter 2 I prepared a study time plan, to try to ensure that I had enough time to complete the work. I also planned to work somewhere that I wouldn't be disturbed.
>
> I found that the study plan never worked out exactly as I expected as there were always interruptions and unexpected events. But the plan was still useful as it allowed me to allocate extra study time to make up for any missed sessions.'

Many people find that they need to combine a plan with some flexibility in this way. However well we plan our lives, the unexpected and unforeseen will always come along. The important thing is to use the plan to help us to get back on track when life makes it difficult to stick to the letter of our carefully thought-out schedules.

The main aim of most agricultural practices is to maximise the amount of energy – or biomass – that is made available in the form of food for us. Less energy is wasted if the food chain is short, with fewer steps. So human beings can get more energy from cereals by eating them directly than they can by feeding the cereals to cattle and eating the meat (see Figure 20). Remember that only about one-tenth of the energy passes to the next step in the food chain.

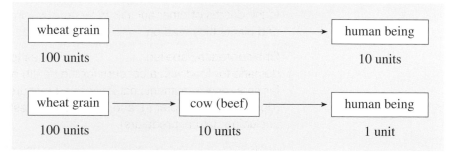

Figure 20   Energy transfer in two agricultural food chains

So, far more food can be produced from a given area of land if plants are fed to people, not to animals. The energy in the beef has passed through two trophic levels by the time it gets to humans, and at each level the amount of energy available to us has been reduced. When we eat plants (cereals, vegetables or fruit) the energy has passed through only one trophic level.

In reality, things are a little bit more complex, as we don't eat all the biomass of the wheat plant (leaves, stalks) or the animal (skin, bones, hooves and so on). And some parts of the world are only suited to growing grass for cattle and livestock, not crops for people to eat. It depends on the plant crop and the animal, and the conditions under which both are produced, but in general arable farming is 20 times more efficient than livestock production, and a meat-based diet requires approximately seven times more land than a plant-based diet.

As the world population grows, larger and larger amounts of food will be needed. One way of increasing the food supply would be to keep food chains as short as possible by decreasing the quantity of meat consumed. But as developing societies progress they tend to move towards eating more meat, not less, with a consequent move up the trophic levels and more demands on ecosystems. The challenge facing us is to develop agricultural techniques that allow us to produce the food we need for a growing population without damaging the ability of ecosystems to continue to feed us in the future.

Activity 25

Allow about 15 minutes

Remembering and forgetting

You'll come across a lot of new ideas – and a lot of new terms – as you work your way through a book like this one. You might be wondering how much you're expected to remember.

You're bound to miss some things on a first read through. That's nothing to worry about. From time to time, look back through a chapter and see how much you remember. Read a paragraph or two at random. Can you still remember the main points? Can you remember what a 'producer' is? What a 'consumer' is? Why is the energy pyramid important, and what does it tell us about our role in ecosystems? Reviewing the big ideas will help to consolidate your understanding.

Look at your notes, and at the terms in your glossary. Do they help? If not, what extra information would make them more meaningful next time?

Comment

It's at this point that you'll find out how effective your housekeeping has been. Can you find the notes you're looking for, and can you find the information you need in those notes? Remember that notes are a way of extending your memory. They help you to remember for longer and to remind yourself of what you have forgotten.

Over time the number of pieces of paper containing notes and diagrams will start to mount up. Devising a system for storing your notes early on will help to prevent problems later. You'll waste a lot of the time you've spent in making your notes if you don't also invest some time in thinking about how you'll make sure that you can find what you need when you need it.

As always, the important thing is to set up a system that will work for you. You are the one who will be consulting your files, and you are the one who needs to be able to find, and understand, the relevant notes. If you think that your system isn't working, especially once you have quite a collection, don't be afraid to go back and change it.

Study checklist

You should now understand that:

- The energy that supports ecosystems is supplied by light from the sun. This energy is used by green plants (producers) to make food. Animals (consumers) obtain the energy they need by eating plants and other animals.

- Decomposers break down dead plants and animals and release important nutrients that are recycled in ecosystems. Energy flows through ecosystems, but the basic chemicals of life are used over and over again.

- Energy is lost along the food chain, from producer to primary consumer to secondary consumer and so on, creating an 'energy pyramid'.

- Human beings use a range of agricultural practices to maximise the amount of energy, in the form of food, that we can obtain from the ecosystems we manage. There are, however, worries about how sustainable some of these approaches may be in the long term.

You should now be able to:

- add terms and definitions to your glossary as you meet new ideas

- plan your study time more effectively

- review your use of time

- organise your notes, so that you can find what you need when you need it.

4 Changing land, changing ecosystems

4.1 Introduction

In Chapter 3, you met the idea of a managed ecosystem, in which, for example, a farmer alters the balance between species to favour the food crop. Humanity has been managing ecosystems for a long time, whether by felling trees or planting crops. Even modern city dwellers often have their own ecosystems to manage at home – their gardens. It could be said that altering the environment in which we live is a basic human activity. Over thousands of years, all this ecosystem alteration has added up to more than a few local fields or gardens – the overall scenery, the landscape, has changed.

Only three tenths of the Earth's surface is land – the rest is ocean. Of the land area, only about a quarter is useful for human activities, as the rest is desert, high mountain, frozen land or otherwise inhospitable. Although this chapter focuses on the UK (a relatively small group of islands in Northern Europe), it illustrates many of the factors that affect land use elsewhere, particularly as increasing populations put more pressure on restricted areas of useful land. Most of the landscape in the UK has been affected by human alterations – there is very little left that is truly wilderness. As an example, Figure 21 shows a photograph taken in the English Lake District. At first glance, this may appear 'natural', but take a closer look. Here, the mountains are natural, but the hedges have been planted to mark field boundaries. Within the fields, there are pastures for grazing and some crops, so most of the land in this photograph is being managed, rather than left wild. Even some of the nearby lakes are artificial, formed by making dams across rivers.

Figure 21 A scene in the Lake District

Activity 26

Allow about 10 minutes   Altered landscapes

Look at the two photographs in Figure 22, showing landscapes that have been altered by people in various ways. How much of the land in each photograph has been altered? Make a note of any evidence you can find in each picture.

(a)

(b)

Figure 22 (a) A city park
(b) A forest

(a) A city park

Almost all the landscape in this photograph has been altered. The river is natural, but the bridge is clearly the result of human activity. Perhaps less obviously, the trees were planted when the park was built. The grass is mown regularly, keeping it tidy, and also preventing the growth of other plants.

(b) A forest

At first glance, this may appear to be natural, but the open path is a hint that people have been at work here. This is a managed woodland – the fairly regular spacing of the trees also suggests that, although it is well established, it is likely to have been planted deliberately.

This chapter will consider how land use has changed, particularly in the UK, and some of the reasons for those changes. As ever, this is a complicated situation, so I will only cover some of the main factors at work here. You have seen how the rest of the food chain depends on the vegetation in a particular area, so it makes sense to focus on the plants, including trees and forests. Along the way, you will see how scientific and technological developments have affected the landscape. Before I describe those changes, there is an important question to answer: what was the British landscape before people altered it? In other words, what is 'natural' vegetation in this part of the world? That question starts a historical discussion that briefly describes some stages in several thousand years of human history.

After that, the story will take a break, so you can practise an important study skill – writing. The environmental story then resumes with the answer to another important question: how does an ecosystem recover from alterations, if left alone? Have you ever wondered what would happen to our parks, gardens and motorways if people stopped using and maintaining them? Once you have considered that query, the chapter ends with an example of an ecosystem that has changed over decades due to the rise and fall of a single species, the wild rabbit. This example shows how interrelationships can have effects which might not be obvious at first. It also illustrates how ecosystem management can require some difficult decisions to balance the needs of different species, including humans.

For now, let's go back to the question I posed earlier.

4.2 What is the natural vegetation of the UK?

To answer this question, let's change perspective to a global view, to see how the UK fits within a wider pattern. Plants, including trees, are very sensitive to temperature and water availability. This may be familiar if you have ever forgotten to water a houseplant, or left it too close to a heater. Even the most vigorous houseplant won't last long if it is kept in the wrong conditions.

Different plants have different needs. Some grow well in warm, moist, rich soil. Others can survive in barren rocky areas, covered by snow for part of the year. Few plants grow in hot, dry deserts, but those that do have adapted to those conditions, with bulbous stems or leaves that store moisture. In contrast, trees adapted to cooler, drier areas have thin, needle-shaped leaves, to reduce the amount of evaporation.

The overall weather conditions in a particular area are known as the climate for that area. For instance, a tropical climate is warm and wet, typically close to the equator in areas with high rainfall. By mapping out the main climate zones, it is possible to categorise the typical vegetation for large areas of the world. These categories are called biomes. As all the other species depend on plants, each biome will have typical insects, animals and other species. For instance, tropical rainforest is a biome that contains particular species of monkeys, parrots and butterflies. None of these species would be found in a desert biome, which is hot and dry, with relatively little plant or animal life.

Globally, there is an overall pattern of about half-a-dozen broad types of vegetation, which divide the biosphere into that number of biomes. Each biome can be classified into more detailed subcategories, which need not concern us in this book. Some examples of global biomes are shown in Table 1.

This table requires some new vocabulary, which you might like to note. Deciduous trees, such as oak, have broader leaves, which they tend to shed in winter. They are more predominant in temperate forest regions. Boreal forest (also called 'taiga') is found in the regions below the Arctic, and its characteristic vegetation is coniferous forest. Coniferous trees, such as pine, are evergreen, with needle-shaped leaves, and produce cones which contain their seeds. Tundra is found near the poles, and at very high altitudes in mountains – in these areas, the ground is almost permanently frozen, so there is very little plant life. Reading across each row of the table shows which biome corresponds with which climate conditions.

Table 1 Biome types and climate conditions	
Biome type	Climate conditions
Desert	Hot and dry
Grassland	Cool/warm/hot, with more water available than desert, but less than forest
Tropical forest	Hot and wet
Temperate forest	Warm, with more water than grassland; mainly deciduous trees, some conifers
Boreal forest	Cooler than temperate forest, with a similar availability of water, although can extend to drier areas with a higher proportion of conifers
Tundra	Very cold (below freezing), fairly dry

Figure 23 Typical trees from different types of forest: (a) temperate (b) boreal (c) tropical

The UK falls within a forest biome (temperate towards the south, boreal towards the north). So, the natural UK vegetation would be mainly deciduous forest in the southern areas with more coniferous forest further north into Scotland. There would be some local variation, as trees tend not to grow above a certain altitude in mountainous areas (too cold), or in marshes (too wet). Even so, trees currently only cover about a tenth of the total UK land area, far less than you would expect for natural vegetation. About 9000 years ago, the UK was almost entirely covered by forest. So what happened to the trees? To understand this, you need to take into account past climate changes, and the history of the UK population. The next section describes some of the major changes in our landscape, and a few of the factors driving them. This is not the whole story, of course: it just highlights some of the developments, mainly in forestry and arable farming.

4.3 The changing UK landscape

In the UK, removal of the forests and the progress of agriculture have profoundly affected the landscape. Currently, about three fifths of the land area is used for farming, including both grazing (pasture) and arable land to grow crops.

Nine thousand years ago, most of the UK landscape was densely covered in woodland, referred to as **wildwood**. The human population was low, probably about 40 000 people in total, less than the population of a small modern town.

Early farmers

Six thousand years ago, **Neolithic (New Stone Age)** farmers were the first to systematically clear large areas of the wildwood, to make space for crops. As the population grew over the next few thousand years, large areas of forest were removed for fuel and building and to increase the area of fields. By 4000 years ago, more than half the original wildwood had already been removed. This may seem remarkable, but it happened over a period of 2000 years. There were probably two main techniques for clearance: felling trees with stone axes (surprisingly effective), and using fire. Forest clearance continued through the Bronze and Iron Ages. By the end of the Iron Age, in the southern parts of Britain most of the remaining woodland was no longer truly wild. It was managed by **coppicing**, which is a way to harvest wood from a tree year after year without killing it. The coppiced tree is cut to a living stump, with enough left to allow the tree to grow new branches. These smaller branches can be harvested continually, giving a regular supply of new wood.

Figure 24   A coppiced tree, showing new branches growing from an old stump

For any farmer, there are some major challenges:

- breaking up the soil for planting
- keeping the soil fertile, as each crop removes nutrients
- protecting the crop from pests, and removing weeds that would compete for resources
- providing sufficient water for the crop.

These challenges have led to the development of new technologies, which have improved as people have found better solutions.

Activity 27

Do this activity as you read the rest of this section

Identifying changes in land use

As you continue to read Section 4.3, make a note of:

(a) the reasons for changes to forests and hedges

(b) other changes in land use.

Comment

Once you have reached the end of Section 4.3, see the comment at the back of the book.

Figure 25   Preparing the ground – modern examples of traditional techniques:
(a) using a version of a digging stick (b) using a simple plough drawn by oxen

To plant seeds, people needed to break up the soil. For the first farmers, the only tool available for cultivating the soil was the digging stick (see Figure 25(a)). The stick is used to break the turf and expose the soil. Using one is slow and very hard work, so productivity is low. At some point farmers started to use animals (oxen and, much later, horses) to pull a simple plough through the soil. This substantially increased the area that could be farmed and made it possible to produce a surplus of food over and above the amount required for the farmer's own needs. Increased leisure became possible and some people could be released from farming to do other things.

Roman influences

Jumping forward in history to about 2000 years ago, after the Roman invasion of Britain, we find a very different situation. The Romans issued Britain's earliest nationwide coins, and the Roman Empire's demand for food and trade goods produced the first full market economy. Ownership of land became more important, as did marking its borders with new hedges. New materials and ideas were brought in from other parts of the Roman Empire, so technology advanced rapidly. The Romans were excellent engineers, constructing well drained straight roads to span the country. Trees were cleared to ensure a good view from the road, and a reduced chance of ambush on a journey. Local clay was fired in kilns to make bricks for building. Quarrying for clay and gravel for building materials altered the shape of the land. Wood was still the major fuel, but by the time of the Romans, only 11 per cent (about one ninth) of the countryside was forest, and very little of that was wildwood.

Boats were the easiest and cheapest means to transport goods throughout the Mediterranean, and within Britain. Sea ports and settlements on the larger rivers became major areas of population. Towns and cities grew with the increasing trade, and the need for crafts as well as food production. It is no accident that London (a large Roman city) is situated on a river that leads to a coast relatively close to continental Europe. As people moved into the cities, more land was used for building.

As the population increased, so did the demand for food for the locals as well as for trade. There is a major problem in farming. If virgin soil is planted for food, yields start to fall after a few years as the nutrients in the soil are removed by the crops. Although plants receive energy from sunlight, and only require water and carbon dioxide for photosynthesis, other nutrients are needed for the plant to grow well. Nitrogen and phosphorous compounds are particularly important. In some parts of the world this drop in yield is overcome by moving on and clearing more land, but clearly it is not practical to do this in a small and crowded country like Britain.

One solution is to rest the land every two or three years by letting it lie **fallow** or unused. Typically, in Roman times, a two-field system would be used, so half the land would be left fallow while the other half was in use. During this fallow period the nutrients in the soil are slowly replaced by the natural processes of **weathering** of soil minerals, which releases nutrients. Stones and small particles of rock are gradually dissolved or broken down by water and frost. Bacteria also take nitrogen from the atmosphere into the soil. This works reasonably well but yields are still modest, especially as only half the available land is covered by crops each year.

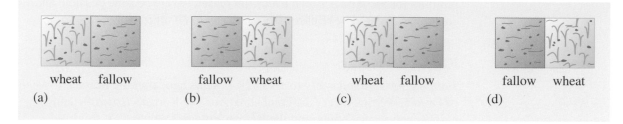

| wheat fallow | fallow wheat | wheat fallow | fallow wheat |
| (a) | (b) | (c) | (d) |

Figure 26   Two-field system of crop rotation: (a) year 1 (b) year 2 (c) year 3 (d) year 4

Yields can be improved further by spreading manure on the land while it is lying fallow, and allowing animals to graze on the stubble of the previous crop. The manure contains many of the nutrients which were removed from the land in the previous crop. By spreading manure on the land, these nutrients are returned to the soil where they can be used again by the next crop. This is similar to the natural processes of nutrient cycling which take place in less intensively managed ecosystems.

From the Norman conquest to the year 1750

Taking another leap forward, about another 1000 years, by the time of the Norman conquest in 1066, about 15 per cent of the land was forest. The Normans kept some of this as a habitat for the newly introduced fallow deer, which they valued for food and the sport of hunting, and thus protected some of the older forest areas. The Normans also brought rabbits to Britain, for use as food, and as a source of fur. As the economy flourished, and the population increased, so did the demand for timber for building and wood for fuel, so by 1400, the land had only 10 per cent woodland. This level of tree cover was maintained for the next few centuries.

At first, most people who worked on the land did not own it, but worked for their 'lord of the manor' under what is broadly termed the 'feudal system'. Each village would have three large open fields, cropped in rotation with wheat, another cereal or left fallow. This three-field system was more productive than the two-field arrangement described above, since less of the land was lying fallow at any one time. Villagers were allocated strips of ground in each field, and also had access to common land for keeping livestock, collecting firewood and so on. But these open areas gradually disappeared as landowners found it more economical to consolidate their land holdings, enclose them with hedges and rent them out to tenant farmers.

Seven hundred years after the Norman conquest, in the early 18th century, a new farming technology was developed – the four-field system of rotation. This innovation depended upon the natural tendency for some plants to increase the nutrients in the soil. The significant discovery was that clover improved the concentration of nitrogen-based nutrients. In a four-field system, each field was used for a cycle of four different crops, in a specific order:

… wheat, clover, turnips, barley, wheat, clover, turnips, barley …

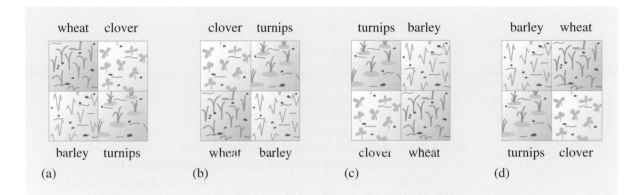

Figure 27   Four-field system of crop rotation: (a) year 1 (b) year 2 (c) year 3 (d) year 4

Growing plants containing nitrogen-fixing bacteria in their roots, such as clover, enabled farms to become much more efficient. The clover restored the nitrogen-based nutrients removed by the wheat and barley crops. Cattle and sheep could be grazed on the clover, providing extra sustenance for the animals, and manure to fertilise the field. In the next year after that, turnips, a root crop, did not depend on the same nutrients as the cereal crops, so a high yield of turnips did not deplete the following two years of barley and wheat. The turnips provided winter feed for the livestock, so that for the first time, a significant number of animals could be kept throughout the year. Before this, most livestock was slaughtered in autumn, as there was insufficient food for the animals in the lean winter months. Experiments were made in selective breeding of livestock to create larger animals producing more meat. Crop production was also made more efficient in the early 1700s by the invention of the mechanical seed drill. This meant that seed could be distributed far more evenly and accurately than the age-old system of scattering it by hand. More effective ploughs were also developed. The resulting increased food supply led to further increases in population.

Allow about 20 minutes   ## Identifying technologies before the year 1750

From Chapter 1, you may recall that technologies are not just about gadgets or machines, they also involve applying knowledge to solve problems, and require people to work together effectively. Technology includes the application of skills, knowledge and tools through good organisation of people.

Go back over what you've read so far in Section 4.3, and make a list of all the technologies you can identify.

Comment

See the comment at the back of the book.

Technologies after the year 1750

By about the year 1750, scientific discoveries in chemistry and physics were beginning to inform new technologies, and over the next 100 years or so there were many changes. So many new technologies and industries sprang up in the late 18th and early 19th centuries that this time is often called the **Industrial Revolution**. By 1850, which is into the Victorian period, changing demands meant that the existing forests were no longer useful. The shift to coal as a major fuel source, for steam engines and steel furnaces, meant that charcoal (made from coppiced wood) was no longer required. Coal mines dominated many areas of the landscape. Timber was imported from abroad, particularly mahogany for furniture. As a result, ancient British woodlands that had survived since about the 1400s were no longer needed. Many of these forests were felled, to be replaced with fast-growing plantations of pine trees, or with agricultural land to feed the rapidly growing population. At the same time, the Victorians planted hundreds of thousands of miles of hawthorn hedges, to mark new patterns of land ownership as the enclosure of open fields was stepped up. Many people found they could no longer find work on the land and drifted to the cities.

Waterways were still a major means of transport – the Victorians built hundreds of miles of canals to transport goods from the industrial centres to the ports and cities. Roads were improved, although horse-drawn carriages were the main means of passenger transport. Once the technologies of steam power and steel production had developed, the first railways were built, with thousands of miles of new track, bridges and tunnels.

The 20th century

In the early 20th century a method of fixing nitrogen from the atmosphere was developed. Artificial nitrogen fertilisers were the result. This was probably the most important farming development in the past 100 years, as it allowed crop yields to increase dramatically. This was a huge change – for the first time in history, people had found a way to add extra nutrients to plant crops, from non-biological sources. It also meant that nutrients were being added from outside the farm, rather than recycled within in it. This has two implications: extra energy (from coal or other sources) is needed to power the chemical process to make the fertiliser; and the amount of fertiliser applied to the land has to be measured carefully – too much can have harmful consequences, as you'll see later.

Horses were used still used to pull ploughs until well into the 20th century. It was not until the 1950s that tractors mostly replaced horses for farm work. Over the last 50 years, tractors have become larger and more sophisticated and their productivity has increased correspondingly. Many hedges were removed between 1950 and 1970, to make larger, more economic fields. More recently, since 1990, more hedgerows and trees have been planted, so that the current land area covered by forest is about the same as in Roman times.

Since 1950, new varieties of the major cereal crops have been developed by systematic selective breeding. These are able to absorb and use very large quantities of nitrogen from the soil. The result has been enormous increases in crop yields – a field of wheat in the 1990s produced more than five times as much grain as the same sized field in the 1940s. These increased crop yields have come at a price. The high rates of use of nitrogen mean that large amounts of artificial fertiliser have to be applied. In the rotational system of farming insect pests were kept in check by the change in crop each year. One benefit of artificial fertiliser was that it freed farmers from the need to rotate crops to maintain soil fertility. It became possible to grow the same crop on the same land for several years at a time, as the lost nutrients were replaced by artificial fertilisers. The disadvantage is that weeds, diseases and insect pests specific to the type of crop are able to build up in the soil as the same crop is grown year in year out, with a break crop every four or five years. This problem is kept in check by using weedkillers (herbicides) and pesticides to protect the crop.

These new technologies have made it possible to feed the increased population in the world. However, the impact on the land, and its ecosystems, has been high. The landscape of lowland Britain has changed as hedges have been removed to make full use of modern farm machinery. Herbicides and pesticides have substantially reduced the populations of wild plants and animals in parts of Britain, although steps are now being taken to remedy this, as you will see later in this chapter.

Excess fertiliser can be washed into rivers by the rain, gradually building up in the water supply. This can have negative effects, for example some water plants can grow more vigorously, out of balance with the rest of the river ecosystem. This can lead to algal blooms – vast floating mats of dense green plant life that can poison and even kill other species. This shows that technology can have both beneficial and damaging effects depending on how it is used. Economic and political considerations have a strong influence on the way technology is applied in a particular region. In developed countries in Europe and North America, government subsidies to farmers can have a profound effect on essential food production, and its environmental impact.

Allow about 20 minutes

Identifying technological changes since the year 1750

Go back over what you've read in Section 4.3, since Activity 27, and make a list of all the technologies you can identify that have changed since the year 1750.

Now write a paragraph, using full sentences, not notes, describing how farming technologies before 1750 compare with more recent ones. Write as though you are explaining this to a friend who has not read this book. Even if you find this a challenge, save your efforts – you will need them for the next section. Note the recommended time for this activity – you only need to produce a first draft at this stage.

Comment

See the comment at the back of the book.

Remember to check your notes from Activity 27 with the comment at the back of the book.

Technology has advanced at an ever-increasing rate since the Victorian era. There have been profound changes in land use, food production, transport and living standards in developed countries since them. The Industrial Revolution was also the start of a trend that has continued ever since – no longer dependent on (or able to be supported by) the land, more and more people are living in cities, supported by food and other goods that are transported over long distances.

4.4 Writing – some basic ground rules

At this point I want to identify some of the key skills involved in good writing in the sciences; I'm going to do this by outlining some basic 'do's and don'ts'. They are basic because other skills (such as organising your writing effectively) depend on them. What I say here relates mainly to the individual small ingredients of writing – the words, phrases and sentences. Think of them as the bricks and mortar that are the building blocks of writing. I'll first outline four hints for good writing. There will be a good number of useful tips and opportunity to practise, which should help you gradually build up your skills and confidence. Good writing takes time to develop, so do not expect to produce a polished piece of work at the first attempt. This particular chapter has been rewritten several times, and then been enhanced by the work of a professional editor. As an author, I know how long it takes to produce good written text!

Allow about 20 minutes   ## The challenges of scientific writing

Writing skills

Track 2 of the DVD is an audio interview with Richard Fortey, who has written many popular science books. He discusses the challenges of scientific writing with Barbara Allen, an OU academic. Listen to the audio, and make a note of the key points you find useful, particularly as Richard discusses various common issues with writing this sort of text.

Comment

Your notes will depend on your own experience. Something that struck me was that Richard does not expect to write more than about 500 words in a single session – worth bearing in mind if you are planning your time for writing, especially as he is an experienced writer.

Hint No. 1: write in proper sentences

It is important to write proper sentences, not just the short phrases you might use in notes. The abbreviated 'Victorians more hedges – boundaries' is an acceptable way to make notes on what you have read, but it isn't a proper sentence. It would be better to write 'The Victorians planted hedges to create boundaries between fields.' The verb 'planted' makes all the difference – you may know that the verb is the 'doing' word. Get into the habit of reading through your work to make sure it sounds right. Reading aloud to yourself, or perhaps to a friend, may help.

Hint No. 2: make each sentence cover a single point

You'll need to be flexible here; two points can be conveyed in a single sentence, but they need to be related in some way. The key part of the advice is not to write sentences that are too long and complex – they are no guarantee that what you are writing is 'good science', or good communication. Simple, short sentences are usually easier to understand.

In the example below, too much is being said in one sentence. You can tell that the sentence is too long if you read it out loud. As we tend to pause for breath at the end of a sentence, you will find that reading a long sentence can become difficult:

> The Victorians took advantage of technologies that used coal, and used steam trains so that many miles of railway needed to be built, with bridges and tunnels, which altered the use of the land.

Better to say:

> The Victorians developed technologies that used coal. Steam trains needed many miles of railway. Rails, bridges and tunnels were built, which altered the use of the land.

Of course, there's an important balance to strike here. A string of very short sentences can be 'bitty' and awkward to read:

> The Victorians developed many technologies. They used coal. Steam trains are an example. These needed many miles of railway. Rails, bridges and tunnels were built. This altered the use of the land.

As a general rule, however, writing in sentences that are too short is less of a problem. It is difficult to understand sentences that are too long, and you can always join two very short sentences together if the writing seems too jerky.

Hint No. 3: write sentences that are concise and clear

Implicit in the need for short sentences is the need to say things in few words. In everyday writing and speech, using more words than we need to is almost second nature, but in science writing it should be avoided.

Look out for phrases that can be abbreviated to one word:

Along the lines of	Like
It being the case that	Because/since
At some future point	Later
Due to the fact that	Because
In the first instance	First

Using too many words, or too many awkward long words, tends to clutter the text and to reduce the impact of what you write.

Which of the following, in general, would you prefer to write:

Enough or sufficient?

Shown or manifested?

Find out or ascertain?

Began or commenced?

Use or utilise?

Need or necessitate?

I hope you feel more comfortable with the short words; they are easier to write, read and understand. Keeping your sentences short is one way to improve your writing. At the same time, think about whether the meaning is clear. This means avoiding ambiguity and saying what you mean.

Hint No. 4: report facts accurately

In science, it is important to check the information you use, and avoid misleading the reader. For example, you need to make sure that you report dates accurately, and that causes and effects are clear. This can be particularly difficult in subjects such as ecology and technology, where there could be multiple causes for a particular effect. For instance, there have been increases in the human population at various times in the UK. In this chapter, these have been linked to improved food production, but there are other causes as well.

Another important writing skill is to make it clear by your choice of words whether what you say is conjecture – how things might be – or fact – how things are. For example, a student wrote 'The Victorians invented artificial fertilisers because they wanted more food.' This is a very intelligent suggestion (and, as it happens, quite close to the truth) but the tutor wrote next to it 'How do you know?' What was needed was a clear indication that this was only a possibility. So don't be afraid to speculate if the occasion is right, but choose your words carefully. This is why phrases such as 'it is possible that', 'it may be' or 'this suggests that' are so frequently used in scientific accounts.

So, distinguishing between facts and possibilities and reporting information correctly are both important for good writing in the sciences. They help give authority to what is written. Some forms of writing require authors to go one step further and give references when they draw on information from other published sources. You can see examples of references in this book. If you are studying as part of a course, your tutor should be able to advise you on this.

Why you need to be your own text editor

Reading through what you write is important. It goes beyond the need for good grammar. It's a way of checking the academic correctness of what you say. Not only should you ask yourself questions about what you've said, you should also reflect on the way in which you've said it. Try to be gently self-critical – this will prompt you to 'edit' your own text. Does that point come across? Is that the right word? Can I say that more concisely? How is that different from what I've said before? Should I say this here or in an earlier paragraph? Asking questions of this type shows you're extending the thinking aspects of writing well beyond the phase of simply putting your first words on paper. Remember your overall aim – to improve the clarity, flow and impact of what you write. Getting rid of all the complicating and unnecessary bits and pieces that clutter your first efforts will identify more clearly the essential message you want to convey. Knowing that you can go back and change what you have written will also reduce the stress of writing. It is far easier to 'have a go' and then change it than to write precisely at the first attempt.

Activity 31

Allow about 45 minutes   Editing your own writing

Now that you have read this section, find your writing from Activity 29 and rewrite it, bearing in mind the advice given above. Then compare your first version with the second one.

Comment

This will depend on your own experience. You probably found that you could improve your work by rewriting it after a break.

4.5 How does a local ecosystem respond to alterations?

Earlier in this chapter, you have seen how we have altered the land for our own use, but what would happen if an area of grassland were left alone?

An ecosystem that is altered or damaged in some way will be out of balance with the biome for that area. For example, if the local biome is forest, but the trees have been removed from one area, then the ecosystem is out of balance. The natural tendency is for plant species to move into that area, bringing the ecosystem back towards the biome state. The spread of a species into a new area is called colonisation. It can happen naturally only if there are ecologically healthy ecosystems nearby to provide plant seeds (for example, sycamore trees produce seeds with a distinctive winged shape – they can be carried by the wind). Once the vegetation has started to recover, insects, birds and other animals will travel into the newly regenerated area.

These processes of ecological colonisation can be supported by environmental management. For example, we are currently seeing important changes in the way agriculture is carried out in Britain. Rather than just maximising food production, farming is becoming more environmentally friendly, with the support of financial subsidies. This new approach increases biological diversity by conserving hedges and the wildflowers, insects, birds and other animals that live on the land. A proportion of agricultural land is left completely uncultivated so that species can gradually colonise it. This provides a habitat for a wider range of species. Leaving some farmland as set-aside is also a way to decrease overall production when that is economically desirable. Note that set-aside land is more permanent than fallow land, which is usually left for only a year. Colonisation is a slow process, taking place over years or even decades.

Bare ground, devoid of all plant and animal life, is a rare sight. A newly erupted volcanic island is one example; a more familiar one might be an abandoned quarry site. In the UK 13 000 years ago, retreating Ice Age glaciers left bare ground. Suppose such land were simply left alone. With time, small organisms such as lichens (a combination of fungus and algae) start to colonise the land. These even grow on bare rock, from spores blown in by the wind. Gradually plant species gain a foothold as seeds and fruits are brought in by wind or water. As plants decompose and rocks are gradually weathered, a soil base develops, which in turn allows new plant species to be established. First insects and then larger animals gradually move in. The important point to remember is that the mixture of species present changes over time in a predictable sequence. The gradual change in the balance of species from a 'standing start' of bare rock is a process called primary succession.

Secondary succession begins on soil cleared of the original vegetation. Clearance can come about by natural forest fires or deliberate removal of vegetation. At first, a few short-lived (i.e. annual) plants become established, completing their life cycle from seed to flowering plant in a single year. Later,

(a)

(b)

Figure 28 Agricultural
ecosystems: (a) a single
species (monoculture) in a
high-yield wheat crop
(b) many wild species in
an area of 'set-aside' at
the edge of a wheat field

a variety of species of grasses become established, including some **perennial**
plants, which have a life cycle of more than a year. Then, small trees and
shrubs (hawthorn, for example) come to dominate the landscape. The number
of different species present changes gradually, and increases. Eventually,
the area will be covered by forest. Your garden lawn would provide a good
example of these processes, assuming you were willing and able to abandon
its care and watch nature reclaim it over a period of 50 to 100 years.

(a)

(b)

(c)

Figure 29   Colonisation of
a paved area: (a) almost
bare paving (b) grasses
and small plants
(c) sycamore seedling
growing nearby

In built-up areas, you can sometimes see colonisation or succession beginning to happen at the edges of paving or tarmac. Figure 29 shows photographs of a paved area that has been left for a few years without attention. Dandelions and grasses are already established, and larger plants are starting to grow in soil that has blown in from neighbouring gardens. Nearby, a sycamore tree seedling has begun to grow among some grasses – Figure 29(c).

If left undisturbed over many years, a woodland containing mature trees develops to the point where few new species of plant appear – the community 'settles' and the balance of species will remain more or less the same once this has happened. This state is called the climax community or climax vegetation, mainly represented by oak woodland in Britain. The dominant tree species may differ depending on the type of soil. Woodland is not inevitable – accidental fires, or severe climatic conditions, may slow or prevent its onset, even if people do not intervene.

Human activity can effectively halt succession in many areas, by preventing colonisation by species we do not consider useful. For example, to a gardener any unwanted plant is labelled a weed, even if it is of a species that would naturally grow in that area (buttercups, for instance, are native British plants). Also, intensive agriculture and uniform forest plantations favour a particular species and alter the habitat. Clearing an area of forest may have other consequences – removing the trees will allow more light to reach the ground, so any dormant seeds may sprout. In different circumstances, the sequence of succession may not always follow the standard pattern – tree seeds may blow in from nearby woodland, so new trees may flourish before smaller plants have a chance to get established.

Because many UK forests were cleared thousands of years ago, what we now recognise as 'natural' or 'wild' grassland must have been maintained in some way. The next section shows how one species – the rabbit – has effectively held back succession in areas of southern England.

4.6   Rabbits – holding back succession in a wild ecosystem

If areas such as the North and South Downs in the UK were to be left undisturbed, what is currently a chalk grassland area would change into mature woodland. This was the original vegetation up to about 3000 years ago, when our Neolithic ancestors cleared the ground for farming.

Nowadays, these chalklands are very rich habitats for many plant species. For example, a number of rare orchids are found there, as well as the early gentian, a plant not found in mainland Europe. The unique plants and animals are partly the result of the presence of rabbits (and increasingly, sheep) that graze the area.

As grazing herbivores, rabbits have preferences for feeding on individual plant species, especially grass, juniper and hawthorn. Other plant species are unpalatable (elder bushes, for instance) and tend to be ignored; such species are likely to increase in number when rabbit numbers increase, because rabbits tend to eat the competing plants.

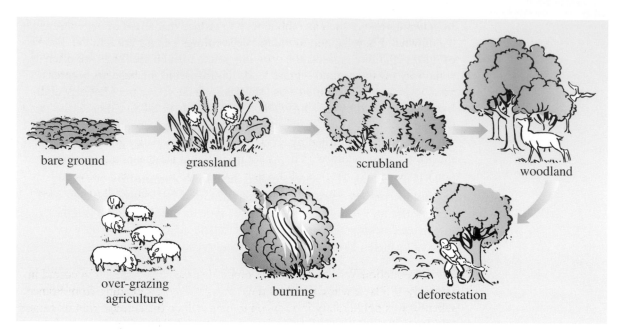

Figure 30   Colonisation of chalk downs in Britain

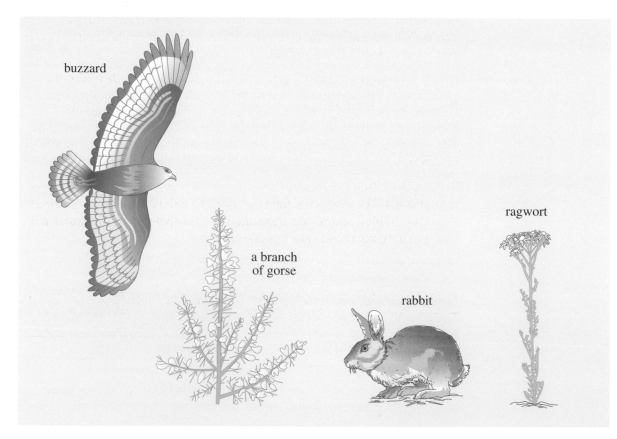

Figure 31   Some downland species

In general, grazers such as rabbits tend to reduce the height of vegetation. Think what might happen to many tree seedlings among grass in the company of grazers. Unlike grasses, the growing points of such seedlings are often some way off the ground – grass tends to grow from the base up, whereas young trees grow from their tips. This means that young trees are especially likely to be killed by grazing rabbits, so areas where rabbits are plentiful tend to have few trees. The activity of rabbits also creates gaps in the vegetation where the soil is disturbed by scratching. This bare chalky ground helps the growth of ragwort, which is a common South Downs plant. If rabbits don't provide bare patches of this sort, other competing plants move in and monopolise the area, and the ragwort seeds find it difficult to establish themselves.

Rabbits and myxomatosis

The introduction into the UK of the rabbit disease myxomatosis occurred in the early 1950s. It was caused by the myxoma virus, spreading from France where it was deliberately introduced to help reduce the damage rabbits caused to farm crops. In some rabbit species, notably *Sylvilagus brasiliensis*, the Brazilian rabbit, the myxoma virus is widespread in the population but has little effect on the rabbit 'host'. But when the European rabbit *Oryctolagus cuniculus* is exposed to the identical virus, it has dramatic and usually fatal effects. The virus is spread by the rabbit flea, which can readily hop from one individual to another.

By 1954 the virus was infecting rabbits throughout the UK. Officially, the government of the day did its best to eradicate the disease. Its efforts were thwarted by farmers who deliberately spread the virus – many farmers transported rabbits which had recently died of the disease to their own lands. The effect on the rabbit population was devastating. Eventually the epidemics of myxomatosis petered out as rabbits developed a natural immunity.

In the mid-1950s, the numbers of rabbits on the South Downs fell because of myxomatosis. The vegetation (and therefore the animal population) changed as a result. Table 2 records the vegetation in 1954 (before myxomatosis) and 1967 (after myxomatosis) at a hypothetical site.

Table 2 Vegetation before and after myxomatosis at a hypothetical site, sampled in different years	
1954	**1967**
Ragwort plentiful	Ragwort scarce
Elder bushes plentiful	Fewer elders
Trees rare	Small trees established
Juniper rare	Juniper flourishing
40 plant species	15 plant species

What this suggests is that the traditional South Downs vegetation reflects a particular form of plant–animal interrelationship – that of grazing. When there are few rabbits (or sheep) around, the plant community changes. Just as animals compete for food, so plants compete for light. Taller plants with larger leaves will overshadow smaller plants. Thus, previously unfamiliar species start to dominate the landscape. What the grazers are doing is limiting the growth of some species – for example, tree seedlings are not able to flourish. The community is held in a state of suspense, preventing its development into woodland. The added advantage is that a greater diversity of plant types is achieved – the rather surprising implication is that unfamiliar and rather rare plants are preserved by grazing!

As there were fewer rabbits by 1967, other animals were affected, reflecting the complex interrelationships at work. Animals that preyed on rabbits – foxes, badgers and buzzards, for example – fell in numbers. The minotaur beetle has larvae that feed just on the dung pellets of rabbits; their numbers were sharply depleted. Bird populations too were influenced; the stone curlew population suffered because it prefers to live on ground that is closely cropped by rabbit grazing. The wheatear declined too; its favoured nesting place is abandoned rabbit holes. On the other hand, animals which compete with rabbits (notably the brown hare) thrived in the immediate aftermath of myxomatosis, owing to the abundance of food.

Activity 32

Allow about 15 minutes

Summarising interrelationships

Go back over what you've read so far in this chapter, and draw a spray diagram which summarises the interrelationships between rabbits and the other downland wildlife.

One way to do this is to put 'rabbits' at the centre of the diagram, and link them to other organisms around the edge, with arrows showing the interrelationships; for example, rabbit holes are inhabited by wheatears.

Comment
See the comment and diagram at the end of the book.

Nowadays the rabbit population of the UK has largely recovered, although myxomatosis still flares up sporadically. For reasons that aren't fully understood, some rabbits initially proved resistant to myxomatosis; they had what I earlier termed 'natural immunity'. Such individuals flourished and produced offspring that were themselves resistant, whereas rabbits that weren't resistant produced much fewer offspring. The result was that resistant rabbits became increasingly common in an expanding rabbit population.

Natural selection

Charles Darwin noticed that in any species there is some variation between individuals – perhaps one plant or animal is slightly larger than another, or more resistant to disease. When there is a struggle for existence, perhaps due to a shortage of food, or a new disease, some individuals will survive longer and leave more offspring. These individuals are said to have a greater fitness than the others. Here, fitness has a very specific meaning – it means the ability to survive and have offspring. If these characteristics are inherited by the offspring, those offspring will have greater fitness, so more will survive to produce offspring of their own. In successive generations, the characteristics of the population may change, for instance, becoming more resistant to disease. Darwin called this process 'natural selection', and he saw it as one of the main reasons for evolution, which is biological change over time.

In certain locations, rabbits are now thriving to such an extent that they are damaging the environment through extensive soil erosion. This has prompted some authorities to instigate controversial plans for the extensive gassing of rabbit warrens, with the aim of conserving the unique chalkland environment. But overall, rabbit numbers on the Downs are probably still not back to pre-myxomatosis levels, so the traditional plants of the downland are still threatened. To conserve the downland, grazing has to be supplied by human interference. Flocks of sheep are moved from site to site to graze. Humans are involved more directly too – hawthorn trees have had to be actively removed 'by hand'.

Activity 33

Allow about 40 minutes

The difference that rabbits make

Now, you can practise writing for a specific task. Use the data in Table 2 and the information you noted in the last activity to write an account in not more than 300 words, to answer this question:

Comparing the situation on the South Downs in 1954 with 1967, what changed, and why, for the following species?

(a) Elder bushes

(b) Buzzards

(c) Brown hares

Comment

Writing tasks of this type are far from easy. Of course, you need to make what you write interesting. It mustn't be over the head of the average reader; remember you have knowledge that casual readers will not have. Specialist terms will have to be explained. You need to think about the audience. Think how different this task is from writing for a professional audience – an assignment that your tutor may mark, of the type mentioned in the last activity.

This is an example of a similar piece of writing from a student:

> 'Writing an accurate account of the changes in populations as the result of myxomatosis with various factors. Farmers spread the disease myxomatosis by carrying dead rabbits on to their farms, to reduce their numbers and prevent damage to the crops, and the disease was carried from one rabbit to another by the rabbit flea.
>
> The rabbits decreased in 1967 so many species were affected, especially the vegetation and animals that depended on the rabbits for food. As myxomatosis spread, the buzzards were affected as they ate the rabbits, so there were fewer buzzards. There were the same number of elder bushes in 1967 as in 1954 due to the fact that rabbits do not eat elder so that species was unaffected. There were very few brown hares before 1954, but their numbers would have increased as the rabbits declined. Fewer rabbits means more food for the brown hares. So, as a result of this change, there were more hares. This shows some interrelationships in the ecosystem.'

Using the four hints for good writing (Section 4.4), let's look at what this student has written and see how it could be improved.

The student example combines some typical problems. Taking each of the hints in turn:

Hint 1 (write in proper sentences) applies to the first phrase, 'Writing an accurate account of the changes in populations as the result of myxomatosis with various factors.' This doesn't sound right, does it? One possible rewrite would be 'An accurate account of the changes in populations as the result of myxomatosis has to consider various factors.' The verb 'to consider' makes all the difference. Although this is now a sentence, it still isn't clear or concise. It might be better to remove it altogether and start with something that leads into the rest of the answer.

Hint 2 (make each sentence cover a single point) applies to much of this example. The writer has combined many ideas in some long, complicated sentences: 'Farmers spread the disease myxomatosis by carrying dead rabbits on to their farms, to reduce their numbers and prevent damage to the crops, and the disease was carried from one rabbit to another by the rabbit flea.' This sentence would be better split into several smaller ones. There is another point to make here. It is easy to get distracted by ideas, and lose track of the original question. The activity does not ask about how myxomatosis spread – the focus is on the changes in numbers of plants and animals. So, to improve the account, all of that sentence could be deleted, and replaced with something more relevant to the question.

Hint 3 (write sentences that are concise and clear). There are several places where the language is muddled. This sentence is unclear, because it seems to imply that the buzzards caught myxomatosis from the rabbits: 'As myxomatosis spread, the buzzards were affected as they ate the rabbits, so there were fewer buzzards.' Buzzards are not affected by the myxoma virus.

This confusion may not be what the writer intended, but what is written here is not clear. A better version would be 'Buzzards need rabbits for food, so fewer rabbits is likely to mean fewer buzzards.'

Hint 4 (report facts accurately) also applies. There are a few places where errors have crept in. 'There were the same number of elder bushes in 1967 as in 1954 due to the fact that rabbits do not eat elder so that species was unaffected.' This contains a factual error and a misunderstanding of the ecology. Table 2 clearly shows that there were fewer elder bushes in 1967. The rewritten account (see below) gives the correct ecological explanation. There is also a guess in this sentence that doesn't relate to any of the facts presented: 'There were very few brown hares before 1954 …'. We are not told anything about the actual numbers of hares. All that can be said is that the numbers will probably be greater in 1967 than in 1954, for the reasons stated in the rewritten account below.

Activity 34

Allow about 30 minutes

Rewriting a draft

Use the comments about the four hints to rewrite this student's account.

Comment

Here is one possible alternative account. Note that it is far shorter than the maximum of 300 words. This is acceptable for many situations where the instruction is 'no more than XX words', although you should check the details with your tutor if you are studying a specific course.

'Between 1954 and 1967, the rabbit population in the South Downs decreased sharply because of myxomatosis. As rabbits are grazers, their absence altered the balance of vegetation in the area. Grazers tend to reduce the numbers of trees and shrubs. This is because tree seedlings put on new growth from their tips, so grazing destroys new growth. In contrast, grasses grow from the base up, so are less affected by grazing. Rabbits also form the food source for other animals, so animal populations were also affected.

There were fewer elder bushes after myxomatosis, as shown in Table 2. Rabbits do not eat elder, but they do consume other plants that compete with elder. So, once the myxoma virus had reduced the number of rabbits, competing plants could flourish. The elder then had to compete for light and other resources, and became more scarce.

Buzzards need rabbits for food, so fewer rabbits is likely to mean fewer buzzards. There are other species available that might provide a substitute food source, for example, brown hares, so the numbers of buzzards might not decrease so sharply. In contrast, brown hares compete with rabbits for food. With fewer rabbits, the brown hares are likely to thrive and increase in numbers, because more food is available for them.'

We've now seen yet more evidence that interrelationships are at the heart of ecosystems. Change one component and a great deal else is influenced. We've also seen that change is the norm, not the exception.

The examples of interrelationships you've come across show just how susceptible ecosystems are to change. Sometimes the term 'the balance of nature' is used to imply that natural systems have a natural robustness and stability. This is far from true: any alterations (such as disease, newly introduced species, loss of resources) are likely to alter the balance rather than preserve it. Of course, change isn't undesirable in itself: value judgements have to be brought into play when the merits of change are debated. Ecosystem management, including farming, requires decisions based on finance, politics, technology, and the needs of humans and wildlife.

Study checklist

You should now understand that:

- Humans have altered the landscape over thousands of years.
- Each area of the world has a natural vegetation type (biome) depending on local climate (temperature and rainfall), and its ecosystem depends on that vegetation.
- An ecologically healthy ecosystem can recover from damage through a process of succession.
- Technology includes the application of skills, knowledge and tools through the organisation of people.
- Environmental management supports the natural processes of ecological recovery through colonisation and succession, by encouraging conditions that are appropriate for the target species.

You should now be able to:

- note the key points from a section of text
- turn a list of key points into a short piece of writing
- identify writing that is inaccurate or unclear
- use the four hints for good writing to improve what you have written.

5 Populations and biodiversity

5.1 Introduction

In this chapter we'll look at some of the factors that influence the number of individuals in a particular population of organisms, picking up on the notion of the energy pyramid from Chapter 3. These factors are applicable to all populations, including our own – the growing population of human beings.

In addition to having domesticated animals and plants for food production we also harvest wild populations (for example, fish from the sea). Some understanding of how and why these populations rise and fall will allow us to manage them in a more sustainable manner.

Finally, we'll look at the rise of the human population and its effects on biodiversity – the number and variety of the living things in specific ecosystems and the biosphere as a whole.

5.2 Population growth

Populations and communities

In Chapter 2 we noted that ecosystems contain a range of different living organisms – animals, plants, fungi and so on. We also noted the importance of being able to recognise and name the species to be found in an ecosystem. Ecosystems contain living and non-living components, and to understand how they work we must gain some understanding of the interrelationships involved.

As well as knowing what sorts of living things are to be found in an ecosystem, it is useful to know how many of them there are and the way in which the numbers change over time. This is particularly true for the biological resources that we harvest for food – for example, fish stocks. If we are to manage these resources in a sustainable way we must understand the rules that govern the rise and fall of populations.

Population science starts with knowing what species are contained in a given ecosystem and how many individuals of each species there are. If we had the time – and it might take a long time – we could try to identify all the species of plants, insects, birds and so on to be found in a given length of Darwin's hedgerow. We could also try to note how many of each species we had found. (To do the same for the microscopic forms of life would be a truly daunting task!)

A number of living organisms of the *same species* that live in a given *place* at a given *time* is known as a **population**.

Note that we have to include three things in describing a population: the species, the place, and the time. So, for example, we might talk of the population of hawthorn bushes in a specific 100 metres of our hedgerow in the autumn of 2005 – just as we could talk about the population of three-spined sticklebacks in a given pond at a given time, or the population of human beings in China in the year 2007. We have to be clear about the species involved to make sure that we don't confuse organisms that look similar. This is one of the reasons that precise identification and naming are crucial. We have to be clear about the location to make sure that we understand the limits of the population. The population of hawthorn bushes in our particular section of hedgerow might not reflect the population in the hedgerow as a whole, or the population in similar hedgerows in other parts of the country. And we have to be clear about the time, because the numbers of a certain species may change from week to week, month to month or year to year. It is important to know exactly what we are talking about.

But, of course, our hedgerow contains more than just hawthorn bushes. It also contains populations of other sorts of trees and other sorts of plants, as well as populations of all the other living things that can be found in that location.

A number of populations of *different species* that live in a given *place* at a given *time* is known as a **community**. The community of an ecosystem is all the living things that live in that particular ecosystem.

Scientific versus normal use of terms

There is a problem with the word 'community'. In normal, everyday use the term suggests a number of individuals working and living together in harmony or co-operation. So, we talk of 'community relations' or 'community spirit', and we complain that people who do not get on with their neighbours are lacking a 'sense of community'. In most cases, things are very different in the biological world. In nature competition is usually the name of the game and, as we shall see, that competition has its roots in the way in which populations grow over time.

The numbers game

Communities involve competition because of the potential for growth in the numbers of organisms in a given population.

Let us examine a population of mice as an example. In conditions that suit them, a population of mice can double every four or five weeks or so. In the real world, however, populations don't grow indefinitely. Something – for example, a lack of food or some other resource – eventually starts to limit the growth of the population.

So, if two mice found their way to a section of Darwin's hedgerow – and there were no mice there to begin with – the population might increase as shown in Table 3.

Table 3 The growth in a population of mice								
Time in weeks	0	5	10	15	20	25	30	35	40
Number of mice	2	3	8	18	33	46	49	50	50

If we plot these numbers as a graph we get the curve shown in Figure 32. The time is shown along the bottom of the graph; the number of mice at the side. The line shows the growth in the number of mice.

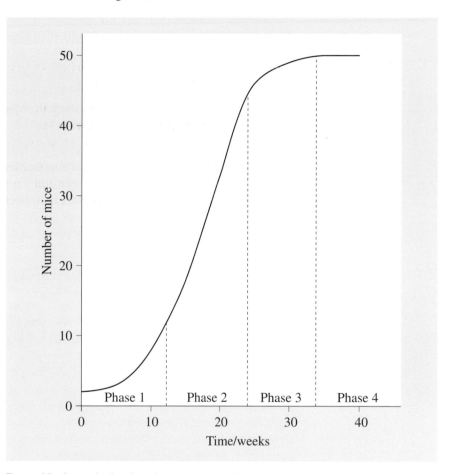

Figure 32   A graph showing the population of mice over time

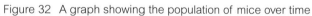

Activity 35

Allow about 10 minutes

Marking a point on a graph

Is it clear to you how the numbers in Table 3 are used to construct the graph in Figure 32? To check your understanding, mark the point on the graph that represents the situation at 20 weeks, when the population is 33 mice.

Comment

See the comment at the back of the book.

We can divide the curve shown in Figure 32 into a number of distinct phases.

Phase 1: The population increases slowly at first, in part because of the time taken for the first few generations to mature to adulthood and start breeding themselves.

Phase 2: The population grows rapidly.

Phase 3: The growth rate begins to slow down as the population starts to be limited by finite supplies of food, water, shelter and other resources.

Phase 4: The population levels off at a constant number.

Most established ecosystems have relatively stable populations, or populations that vary in predictable ways. So there must be physical or biological aspects of the ecosystem that regulate the size of the population. These are what start to slow the growth of the population in phase 3 and bring it to a more or less constant level in phase 4. But what are these limiting factors? Why doesn't the population just continue to grow and grow?

Mice can enter the population of the hedgerow in two ways: birth, or immigration into the hedgerow from another population somewhere else. **Immigration** is the movement of part or all of a population from somewhere else *into* an area. Similarly, there are only two ways in which mice can leave the population: death, or emigration from the hedgerow. **Emigration** is the movement of part or all of a population *out of* an area.

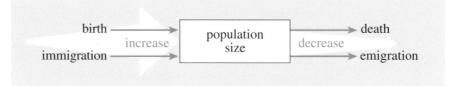

Figure 33   The factors that influence population growth

So the change in the population of mice in an ecosystem at any point in time equals the difference between the number of mice being born and the number dying, plus the difference between the number moving into the ecosystem and the number moving out of it. We can express this as a word equation, which says exactly the same thing using words and the mathematical symbols for minus (–), plus (+) and equals (=):

change in the population = (birth – death) + (immigration – emigration)

Carrying capacity

If our population is steady (phase 4), we know that the number of mice coming into the ecosystem (through birth and/or immigration) must equal the number leaving the ecosystem (through death and/or emigration). So, as the population levels off at a constant number:

birth + immigration = death + emigration

Why does this happen?

In practice, the size of a population is limited by a number of living and non-living factors in the ecosystem. One of the main limiting factors is competition for resources. Animals compete for food, water and shelter. Plants compete for light, water and nutrients. There are finite supplies of all these factors. As the population rises, the resources start to run out. Competition for food, space and so on becomes more intense. Death and/or emigration rise to match the birth rate and the population stabilises. (Similar rules apply to plant populations, although the dispersal of seeds and the success of seedlings in some places and the failure of seedlings in others replace immigration and emigration.)

The maximum population size that can be maintained in a given ecosystem is called its **carrying capacity**.

Competition for resources is not the only factor that can influence the size of a population. As a population of organisms grows other factors come into play, and some of these are associated with other populations in the community: for example, increases in infectious diseases caused by microscopic organisms, increases in the numbers of parasites, and increases in the numbers of predators. In most cases, the population of any given species in an ecosystem is regulated by a whole network of interrelationships between the species and other living and non-living components of the ecosystem.

5.3 Interrelationships between populations

Populations don't exist in isolation. The growth, or decline, of a given population is determined by its interactions with other populations in the ecosystem, as well as by the limitations imposed by the physical resources available.

Some populations go through repeated and regular periods of population peaks followed by population falls. One of the best-known examples of these 'boom and bust' populations is provided by the Canadian lynx (scientific name *Lynx canadensis*) and the snowshoe hare (scientific name *Lepus americanus*).

Figure 34 The food chain in action: a Canadian lynx chases a snowshoe hare

The Hudson's Bay Company controlled the fur trade in North America for several centuries, and an analysis of its records for a period of more than 100 years proved what the Canadian fur trappers had claimed all along: the numbers of lynx and hare fluctuated over time in a regular series of cycles. Figure 35 shows the changes in the population sizes of lynx and hare over a sample period of 90 years.

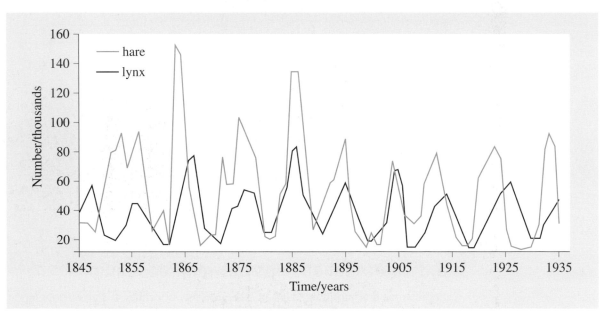

Figure 35   A graph showing the populations of lynx and hare over time   Source: MacLulich, 1937

Allow about 10 minutes   Reading information from a graph

Note the positions of the peaks of both populations.

What is the approximate gap between the peaks of the lynx population?

What is the approximate gap between the peaks of the hare population?

Do the peaks of the lynx population occur at the same time as the peaks of the hare population? If they don't, are the lynx peaks before or after the hare peaks?

Comment

The peaks of both populations are roughly 10 years apart, but the peaks of the lynx population tend to occur slightly later than the peaks of the hare population.

The hare population was found to rise and fall in an approximately 10-year cycle, with that of the lynx following about two years behind. The hare is an important source of food for the lynx, which is why there is such a clear link – a clear interrelationship – between the two populations.

When food is plentiful more of the hares' offspring survive and the population increases rapidly. But as the population gets larger and larger the hares eat more and more of their food supply and start to approach the carrying capacity of the ecosystem as far as hares are concerned. Eventually, shortage of food means that the hare population begins to fall.

The cycle in the hare population is matched by a cycle in the lynx population. When the hare population is growing, the size of the lynx population increases in response. (Remember the energy pyramid from Chapter 3. The numbers of lynx depend in part on the numbers of hares available as food.) The lynx population responds to the increase in their food supply by raising more kittens.

But the more lynx there are, the more hares will be eaten – especially as the hungry hares are forced out into the open in their search for food. The population of hares begins to fall, followed by the population of lynx as their food becomes more difficult to find. The plants that form the food of the hare recover … and the cycle begins again, as shown in Figure 36.

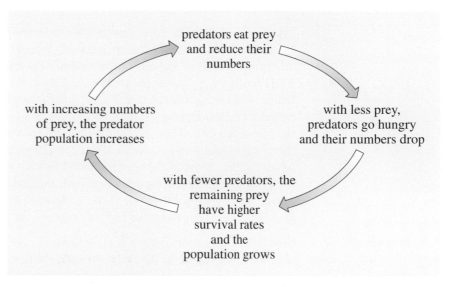

Figure 36   How populations may cycle over time

This predator–prey relationship is unusual in that the hares form a large part of the diet of the lynx, so the fates of the two populations are intimately related. But even so, there are lots of complicating factors. The populations of other participants in the drama – grouse, squirrels, ravens and so on – rise and fall alongside the populations of lynx and hare. In most communities, populations are regulated by a number of interrelationships involving other living organisms (as food, predators, parasites and so on) as well as non-living components of the ecosystem (for example, climate, shelter, available water and so on).

5.4 Managed populations

The biology of populations has a number of practical consequences for human beings. As well as farming plants and animals for food, we also harvest wild populations of plants and reduce animal populations through fishing and hunting.

An understanding of how populations grow should allow us to determine the maximum sustainable yield (MSY) for a given species in a given ecosystem: that is, how many individuals we can remove in a given time period without damaging the population's ability to replenish itself. If we take more than the maximum sustainable yield – more than the population is able to replace through birth or immigration – we run the risk of the numbers falling. In some cases, this can eventually lead to the collapse of the population and, in extreme cases, the loss of the species from the ecosystem altogether.

An example may make all this a bit clearer.

The seas off the east coast of Newfoundland in Canada were once so full of cod (scientific name *Gadhus morhua*) that huge shoals of fish almost prevented the progress of the explorer John Cabot's ship when he sailed to the region in 1497. The sailors could catch the fish for food by simply scooping them up in baskets.

The notion of exceeding the maximum sustainable yield must have seemed laughable when the first fisheries were set up several hundred years ago. After all, a female cod can lay up to nine million eggs a year and live for 20 years. The French writer Alexandre Dumas, writing in the 19th century, calculated that if all these eggs hatched and grew into mature fish, and these fish in turn raised all their offspring to adulthood, it would take only a few years to fill the sea so full of cod that it would be possible to walk across the Atlantic on their backs from North America to Europe. (In reality, of course, this could never happen. Most of the eggs and small fish are eaten by other fish, and the growing cod themselves compete with each other for food and other resources. Remember the mice in our hedgerow: a population's potential to grow is always limited by the carrying capacity of the ecosystem.)

Some 500 years after Cabot's visit, in the early 1990s, decades of over-fishing resulted in a collapse of the cod stocks and the devastation of the marine ecosystem. More efficient fishing technologies, and increased demands for fish from a growing human population on both sides of the Atlantic, had reduced the population to a tiny fraction of its original size.

There were warnings. Scientists had long pointed out that catches exceeded the maximum sustainable yield and that the population of cod was in decline. But policy makers had to balance their advice against the needs of the fishing industry and the communities it supported. 'Draggers', huge nets as long as a football pitch, continued to be pulled along the sea floor, catching mature, breeding cod as well as removing everything else that acted as food and shelter for young fish. In 1992 it was finally realised that the cod population had collapsed and the Canadian government closed the fisheries. Over 40 000 people lost their livelihoods.

To try to prevent this happening again, politicians and scientists have been working together to find new ways of managing and harvesting fish populations in a sustainable manner. An example of this new trend is provided by the Ecosystem-Based Management (EBF) approach. This approach recognises that healthy ecosystems are necessary for healthy populations of fish, so it involves restoring those ecosystems as well as other measures to improve the population levels of the harvested species. It includes 'resting' endangered species to allow their populations to recover; closing certain areas to fishing to protect vital feeding, breeding or nursery areas; and changing fishing practices to protect the marine environment in general.

In this way, it is hoped that cod and other commercially fished species will breed successfully and the young fish will be able to find the food and shelter they need to mature. The populations can then increase – and be managed – to support controlled human harvesting at, or below, the maximum sustainable yield.

The Devil's gardens

Human beings are not the only species to manage populations and communities for their own purposes.

The Amazonian rainforest is probably the most diverse ecosystem on the planet. It is thought to contain over half the world's known species of plants and animals, and it is anyone's guess how many species remain to be discovered.

Most of the rainforest is made up of complex communities that contain many species of trees and other plants. But deep within the forest are large areas dominated by a single species of tree: scientific name *Duroia hirsuta*. These areas appear to occur at random, and for many years scientists were at a loss to explain how they were created. The local inhabitants believed that these gardens were tended by evil forest spirits, hence the name 'the Devil's gardens'.

Recent research has shown that the trees are home to a species of ant, scientific name *Myrmelachista schumanni*, which nests in the hollow stems of the twigs. The ants poison other plants by injecting them with formic acid to ensure that there are plenty of Duroia trees to hold their nests. The 'gardens' may last for up to 800 years and at any one time can be the home of up to three million worker ants and 15 000 queens.

5.5 Using your scientific judgement

You'll appreciate by now that learning science is not simply a process of passively taking in facts and figures. You are faced with decisions – choices and dilemmas, such as how long to study, which topics to go over again, and which topics are the most important. And increasingly the science you learn will raise a different, though equally taxing, set of questions. These are a mixture of moral and practical problems: how the science you learn should be thought about, how it should be applied, and how it affects other people.

With a topic such as the environment, wrestling with these 'bigger' questions forms part of the learning process. You have to learn to think about science in ways that help you to come to personal decisions. What study techniques can you use when 'big questions' arise? Is the evidence for this particular argument strong or weak? Is that approach to conservation justified? What follows is one example of personal decision-making, rooted in science, and based on a genuine contemporary environmental problem.

The conservation of red grouse

This section involves listening to Track 3 on the DVD. It is about the shooting of red grouse, a game bird about the size of a small chicken which is found exclusively in the upland moors of northern and western Britain. In recent years, about 400 000 grouse have been shot each year, generating a gross income of about £10 million.

hen harrier

a branch of heather

red grouse

Figure 37   Red grouse (scientific name *Lagopus lagopus*) and other moorland species

You'll hear a variety of points of view on the DVD. To help you make sense of them, you'll need some background to the issues.

A dilemma for conservationists

1   A grouse moor is an artificial ecosystem, and it has a distinctive mixture of plants and animals. The dominant plant is heather, which provides cover for grouse and is the major source of their food.

2   If the habitat were simply left alone, it would gradually change: grasses would take hold and shrubs and small trees would become more of a feature. Such a habitat would no longer be able to support large numbers of grouse.

3   The grouse moorland habitat is managed by gamekeepers to maximise the numbers of red grouse for shooting. Animals that feed on red grouse or their eggs are often shot – foxes and crows, for example.

4   Patches of heather are occasionally burned to encourage the growth of new heather shoots, which are an important food source for young grouse.

5   Managed moorland of this type attracts a wide variety of animals and plants that thrive in this particular habitat. For example, meadow pipits are common birds on heather moorland.

6   Red grouse moorland has attracted increasing numbers of rare birds of prey. Such birds are called **raptors** and their diet includes grouse, especially the young and the eggs. In particular, numbers of hen harriers had increased significantly in such moorland in the years before this audio track was recorded.

7   At one time, the hen harrier was a relatively common bird throughout the UK, but its numbers declined sharply, largely because they were persecuted. Hen harriers are now protected by law; it is illegal to shoot the birds or to interfere with their breeding, for example by disturbing their nests.

8   The numbers of grouse in managed moorland habitat have been in overall decline for the past 30 years or more. The ecological reasons for this long-term decrease are complex, and moorland has not always been well managed. For example, increased grazing by sheep discourages the growth of heather. Grouse numbers are also adversely affected by disease.

9   Apart from the long-term decline, the numbers of red grouse vary in the short term. There are cycles of population increase and decrease (rather like the lynx and hare pattern in Figure 35), and the time from peak to peak is typically around five years. The causes of these population cycles are not known, but they are not believed to be related to shooting.

10 Because the numbers of raptors (especially hen harriers) were increasing at a time when grouse numbers were decreasing, those who managed grouse moors became convinced that the hen harriers were particularly to blame for poor 'bags' of grouse in recent shoots.

11 Gamekeepers were prevented from interfering with hen harriers because of European Union conservation laws. Some felt that the numbers of hen harriers should be limited to a set 'quota' and that excess birds should be removed to other sites. Conservationists opposed such moves, on the grounds that they would be ineffective and illegal.

12 A number of concerned organisations agreed to support a five-year study on the relationship between hen harriers and grouse numbers at Langholm Moor in south-west Scotland. The report concluded that the large numbers of raptors present contributed to the low numbers of grouse, though habitat mismanagement was very important too. Shortly after the report was published in 1998, Langholm ended grouse shooting, as the following extract from *The Times* reveals.

An inglorious ending for Churchill's grouse moor

By Shirley English

ONE of the country's best-known red grouse moors, and a favourite of Churchill's, will no longer play host to guns because there are not enough birds.

The decision to end grouse shooting indefinitely at the 12,000-acre Langholm Moor in the Borders, which is part of the Duke of Buccleuch's estate, was attributed to the high concentration of protected birds of prey, which hunt the grouse for food.

Langholm Moor volunteered in 1992 to be the site of a five-year study by Scottish Natural Heritage, the government environmental advisers, into the conflict between protecting raptors and sustaining viable grouse populations.

Shocked landowners, conservationists, scientists and government advisers were told of the closure yesterday during a Moorland Summit in Perthshire on the crisis facing many of Britain's 400 grouse moors, in which they were to be told the successful results of tests carried out on the estate, involving the feeding of dead rats and mice to raptors in an effort to wean them off grouse.

Langholm still holds the Scottish record for the largest grouse bag when, in 1911, 2,523 brace [pairs] were shot in a day. Now there are just 1,000 grouse left on the whole estate.

In 1990 the estate boasted grouse bags of more than 4,000 in the season. Last year it fell to 51 bags.

To break even the estate needed to bag 1,100 brace (two grouse worth about £70) a year.

Making the announcement, Gareth Lewis, the factor for Buccleuch Estates, said the moor, which costs £100,000 a year to maintain, was no longer economically viable.

Source: The Times, 1998, p.3

Soon you will be asked to listen to Track 3 of the DVD, but first some guidance. Early on you will hear a variety of views presented, some of which represent interested organisations. It's not important that you remember exactly who says what, or the details of the organisations they represent, but it might be useful to list the participants and their organisations before you listen.

> Pippa Greenwood, horticulturalist and broadcaster
>
> Ian Newton, the British Ecological Society
>
> Iain Bainbridge, the Royal Society for the Protection of Birds
>
> Dick Potts, Director General of the Game Conservancy Trust
>
> Des Thompson, Scottish Natural Heritage
>
> Gareth Lewis, Factor (manager) of the Buccleuch Estate, which includes Langholm Moor

These initial contributions from the experts last for about 10 minutes. Later, Joan Solomon, an OU academic, played this recording to three prospective students and recorded their subsequent discussion. None of these brave student volunteers had any special knowledge of the subject, although they were supplied with the background information that you've just read. She was interested in their reactions. What sense could they make of what they heard? When you listen to the students' discussion on the DVD, you can compare their reactions to the 'experts' with your own thoughts. What she wanted to find out was how the students responded to controversy, and the sometimes bewildering variety of conflicting opinions.

Activity 37

Allow about 60 minutes

Listening and disagreeing

The red grouse debate

Now listen to Track 3 on the DVD. As you listen to the expert contributions, make notes to help you make some sense of them. Listen again to any parts that you are unsure about. Then write down your own views about whether shooting grouse is justified.

Now listen to the students' discussion. Note how their reactions compare with your own. Do you think they took the experts' views on board? Finally, listen to Dr Jeff Thomas's thoughts at the end of the track, where he reflects on what he feels can be learned from the exercise. Write down what you feel you learned from listening.

Comment

This audio track was recorded in 1998. Ten years later, Langholm Moor is still the subject of controversy – and the local ecosystem does not support grouse or hen harriers in any significant numbers. Other grouse moors in Scotland are faring better. In a controversial case such as this, scientific information can act as a guide to your thinking. But your views on other issues are vital too – on animal rights, for example, or the importance of commercial interests, on how convincing you find the arguments for conservation, and so on.

Whatever decision you come to, you ought to be able to offer reasons to support it. That way, decisions are more informed and considered – and you can feel more confident about your point of view. Becoming confident about the science you know and of how it should or shouldn't be applied is an integral part of being a successful student. The fact that there is so much more to find out isn't a cause for anxiety – it's a reason for sustained interest. With luck, it's an interest that can, and should, last a lifetime.

5.6 Human population growth

There is one species whose population has seen a remarkable growth. Figure 38 shows the growth in the population of human beings. '(B)CE' stands for '(Before) Common Era', which is tending to replace 'BC' and 'AD'.

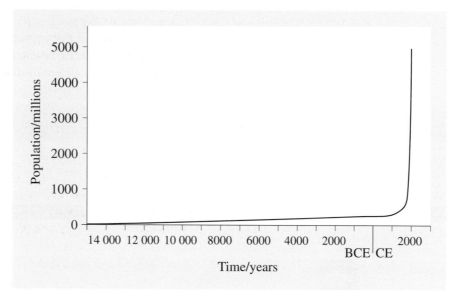

Figure 38 The population of human beings over time

Allow about 5 minutes

Comparing graphs

Look at the graph in Figure 38. Which of the four stages from Figure 32 does the present population growth represent?

Comment

Human population growth is still in phase 2: rapid increase.

The population of human beings continues to increase because of our ability to increase the efficiency with which we exploit the ecosystems we manage for food and other resources. Our agriculture allows us to increase the amount of energy available to us as food, and the spread of human populations across the biosphere demonstrates our ability to migrate from one ecosystem to another in search of new resources. Since the 19th century in the industrialised West, improved hygiene and the development of medicine have also led to fewer infants dying, and people living longer.

But population growth and industrialisation increase our demands on the biological and physical resources of the biosphere, and we are still dependent on the carrying capacity of the ecosystems we inhabit. In some parts of the world, the brutal reality of the energy pyramid means that human populations remain level, or even fall, as people die because they do not have enough to eat.

It is worth remembering that there is a limit for all of us: the carrying capacity of the biosphere as a whole. The energy pyramid operates at a global level, and plants remain essential for the food we eat (and the oxygen we breathe). And at present, a lot of our other energy needs, apart from food, are provided from finite sources of fossil fuels. As the human population nears the carrying capacity of the biosphere, the competition for resources (water, food, oil) and the consequences of overcrowding (pollution, disease) will tend to increase.

There is some good news, however, in that figures from the United Nations Population Fund have indicated that the rate of the rise in the human population may be levelling off (see the dashed line on Figure 39). One reason may be that improved health care, including reproductive health, has led to people choosing to have smaller families. (In many parts of the world, high mortality rates, including high infant mortality, mean that people have large families to ensure that enough children survive to care for their parents in their old age.) A far less positive reason may be the increasing mortality rates in sub-Saharan Africa and parts of the Indian subcontinent, in part due to HIV/ AIDS.

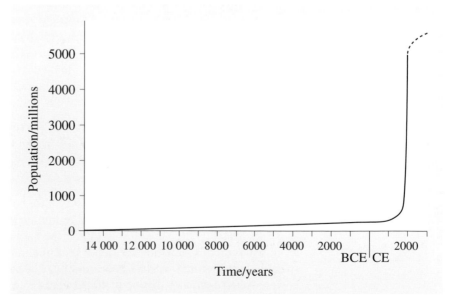

Figure 39 The population of human beings over time, showing the predicted trend

Plague

Infections are able to pass from host to host more easily as the population density of the host increases. For this reason, widespread outbreaks of infectious diseases (epidemics) among human beings are particularly severe in cities. In fact, for most of the period since humans began living in cities, city populations have been maintained only through continual immigration from the countryside. Not until the development of community sanitation, immunisation and other public health measures did cities avoid periodic sharp drops in population as a result of epidemics.

The recurrent epidemics of the 'Black Death' in Europe that began in the 14th century caused a sharp decline in the population. In just five years (1347–1351) at least a quarter of the population of Europe died from the disease (probably bubonic plague).

More recently, the great influenza (flu) epidemic of 1918–1919 is thought to have killed somewhere between 20 and 40 million people worldwide. More people died of influenza in a single year than in the five-year period of the Black Death, and the entire epidemic killed more people than World War I (which ended in 1918).

5.7 Biodiversity

In Chapter 2 I started with the biosphere as a whole (the 'pale blue dot' of Chapter 1). I then worked my way down, through ecosystems, to look at the interrelationships between organisms, and the interrelationships between organisms and the environments in which they live. In this section I'd like to reverse that process, and move from populations and communities at the ecosystem level to examine the numbers of species at the level of the biosphere as a whole. This is because the growth of the human population is resulting in the loss of species from ecosystems. The problem is not the loss of the odd species here or there in individual ecosystems, but the loss of large numbers of species from ecosystems across the world. This makes species loss a global, as well as a local, issue.

A term often used in connection with the complexity of life on Earth is biodiversity (short for 'biological diversity'). The word 'biodiversity' can be used to refer to the amount and variety of life on Earth as a whole, or to the amount and variety of life to be found in a particular ecosystem (for example, the 'biodiversity of the Amazonian rainforest').

It makes sense to look at biodiversity after human population growth because, as we shall see, the two are closely linked. As the human population grows, as human societies become ever richer in economic terms, biodiversity – the 'wealth' of the natural world – is decreasing. The science magazine *Nature* has published estimates that up to 40 per cent of known land-based species could be 'committed to extinction' by the year 2050 – which is 44 years away as I write this sentence (Thomas *et al.*, 2004). It is hard to be certain about the numbers. We don't know how many species there are, and we don't know how quickly they are being lost, especially as most estimates concentrate on the larger animals and plants and ignore the small and the microscopic. But whatever the exact figures, there seems to be little doubt that the world is experiencing a crisis in terms of species loss. Plants and animals are being lost from the biosphere at such a rate that many scientists are talking of a 'mass extinction', similar in scale to the one that ended the reign of the dinosaurs.

Biologists and conservationists use the acronym HIPPO to summarise the main threats to biodiversity. You'll probably recognise many of these threats as consequences of the way human beings modify ecosystems to obtain biological and physical resources.

H is for habitat change. We have already examined the close links between populations and communities and the physical environments they inhabit. Change the habitat – by cutting down trees, say, or draining the water – and some species will disappear from that location. And surely no species has ever changed the face of the world in the way that we have.

I is for invasive species. As human beings travel the globe they take plants and animals with them. These organisms find themselves in new ecosystems, without the interrelationships and limiting factors – for example, predators – that would otherwise tend to restrict their populations. The results can be devastating, as we shall see later in this chapter.

P is for pollution. This is a form of habitat change, but the by-products of human activity are such a threat to ecosystems and the living organisms they contain that they merit a category all of their own. Pollution affects land, sea and atmosphere, as you will see later in this book.

P is for (human) population growth. The more of us there are, the more we demand from the ecosystems around us in terms of food, water and other resources.

O is for over-harvesting. We harvest plants and animals for food, and other uses such as timber, but if we take more than their populations can replace, we damage the ecosystem's future ability to sustain us and the other living organisms that depend on those plants and animals.

In looking through this list, you may have noted that human population growth makes all the other factors worse. The World Conservation Union (IUCN) produces the Red List of Threatened Species, which lists the organisms known to be threatened by extinction. Of more than 16 000 at-risk species in the 2006 list, over 99 per cent were threatened as a consequence of human activity. The main problems were habitat loss and habitat degradation, which threatened over 80 per cent of the endangered mammals, birds and amphibians. The other HIPPO factors – invasive species, pollution and over-harvesting – also played an important role in some cases.

As human beings use more and more of the Earth's resources, we find ourselves in a life-and-death struggle with the rest of the living world. In the short term, our numbers and technology mean that we prevail in most of the local skirmishes, but in the long term this is a contest with no winners. We, and all other living things, remain dependent on the ecological health of the ecosystems we inhabit.

Goodbye 'Turgie'

Conservation is difficult. A simple, low-key example should make some of the HIPPO issues a little clearer. I could have picked any one of a number of more famous, high-profile extinctions or large-scale collapses in biodiversity from ecosystems across the world. But this example shows how easy it is to overlook the complexities of the interrelationships that are fundamental to ecological health.

Hawaii could, with some justification, be called the 'extinction capital of the world'. Many of the islands' plants and animals evolved in isolation from other ecosystems, in a number of diverse habitats from upland forests to the coral reefs that fringe the islands.

Habitat change started with the first Polynesian settlers to arrive in the islands. Now some three-quarters of the original forests and grasslands have been cleared for agriculture and development. The increase in the human population has caused a number of problems in addition to habitat change, including pollution and over-harvesting.

The first Polynesian seafarers also brought a number of invasive species with them, including pigs and rats. Later settlers added cats, goats and cattle. Other species have been introduced by later waves of immigrants, tourists and, to their shame, even some scientists.

In the early 1900s some Hawaiians introduced the African giant land snail (scientific name *Achatina fulica*) to the islands as a pet and a garden ornament. The snails soon escaped and their numbers expanded, causing havoc as they chomped their way through native plants. In what must have seemed like a good idea at the time, scientists decided to introduce yet another species of snail to control the numbers of the African giant land snail: the predatory rosy wolfsnail from the south-eastern USA (scientific name *Euglandina rosea*). The idea was that the wolfsnail would reduce the numbers of the African giant to manageable proportions by the simple tactic of eating them.

Unfortunately, the rosy wolfsnail ignored its intended target and started to eat the already endangered native Hawaiian snails. To make things worse, at some point it found its way to the neighbouring island of Moorea and devastated the local snail populations there as well. Of seven Moorea species of *Partulina* snail, six have been saved from extinction only by breeding them in captivity.

At 5.30 pm on 1 January 1996, the last surviving member of the seventh species, scientific name *Partulina turgida*, died of an infection at London Zoo – some 10 years after the species had become extinct in the wild.

In memory of 'Turgie' and his species, the staff prepared a little memorial:

Partulina turgida: 1.5 million years BC to January 1996

Is biodiversity important?

The loss of living organisms from the biosphere worries and dismays many biologists, naturalists and environmentalists. But should it matter to the rest of us? The death of Turgie is bad news for Turgie, for the species *Partulina turgida*, and for the keepers at London Zoo who worked so hard to save it. But is it something we should worry about? After all, I'd be very surprised if you'd ever heard of Turgie, or *Partulina turgida*, before you started this chapter. There were seven species of *Partulina* snail; now there are six. So what?

Rather like the issue of grouse management, the answers to questions of this sort depend on a mixture of objective scientific observations and subjective value judgements about the kind of world we want to create for ourselves and our children.

Activity 39

Allow about 5 minutes

What does 'biodiversity' mean to you?

Before reading on, stop and ask yourself what biodiversity means to you. Do you think it is important? Why? Try to jot down two or three points.

Comment

There is no one right answer to a question like this. In part, your answer will depend on your own values and priorities. No one can tell you what your attitude should be, although I hope that you'll find some of your opinions mirrored in what follows.

First, there are good pragmatic reasons for valuing biodiversity. A lot of the medicines we use were discovered in, and are produced from, plants and other living organisms. Losing species at such a rate means that we are losing medicines and treatments we don't yet know exist, as well as potential crops and other resources. Living things also provide us with a range of services that we take for granted but would find it difficult, if not impossible, to replicate by other means (for example, many of the crops we grow depend on pollination by insects).

It also seems that biodiversity is important for – as well as being a measure of – the health of ecosystems. As we have already seen, living organisms create the rich soils that plants grow in as well as the air that we and other animals breathe. We depend on ecosystems, and ecosystems depend on the interrelationships between green plants, bacteria and innumerable other organisms in ways that we don't yet fully understand.

As species are lost, some of these interrelationships are lost, and some of the network of links that connect the living components to each other and to their environment are severed. Ecosystems vary in their ability to resist change (ecological resistance), and in their ability to recover from change (ecological resilience), and it appears that biodiversity plays an important role in both. In general, the more complex the 'meshwork' is – the more species and interrelationships there are – the better the ecosystem can cope with the loss of individual links. However, observations of how change affects ecosystems in practice have indicated that all species may not be equal in this respect. The loss of some species – sometimes called **keystone species** – can have a drastic effect on the health of an ecosystem. The reasons are not always clear, and they seem to vary from case to case, but it seems that the loss of certain species has the capacity to cause drastic changes to the ecosystem as a whole.

Successful conservation

It's sometimes tempting to become a little despondent in the face of all this gloom, so it's reassuring to be able to report that there is good news out there as well.

The complexity of the interrelationships involved in ecosystems means that a species needs other living things and a specific physical environment to survive. To protect a species, you have to protect an ecosystem. And this is precisely what many of the most successful attempts at conservation have set out to do.

The first nature reserve, Yellowstone National Park in the USA, opened in 1872. There are now over 40 000 wildlife reserves across the world, covering an area equivalent to India and China combined. The world's largest reserve, Tumucumaque National Park in Brazil, covers an area the size of Switzerland and is just one of a number of 'mega-reserves' that form part of the Amazon Region Protected Areas programme (ARPA). Preserving habitats on this scale will save thousands of species we know about … and millions that remain unknown to science.

Other parts of the world are introducing similar initiatives. In 2002 the Chinese government announced an ambitious plan to cover 5 per cent of the country with new forests, including reserves for the giant panda and other endangered wildlife.

These practical, resource-based arguments for conservation are compelling. For me, however, there are other, subtler, more personal reasons for mourning the decline in biodiversity. The world is a complex and beautiful place, and I think it is a privilege to be part of it. The loss of a species – like the loss of a human language, or a human culture – diminishes the beauty of the world simply by removing a little of that complexity.

The world I will leave for my children will be poorer for the loss of *Partulina turgida*, just as it would be poorer for the loss of more glamorous and threatened species such as the tiger or the orang-utan.

The message of ecology is that we are part of a network of links, of interrelationships, between living things and the environment, at the local, ecosystem level and at the global, biosphere level. We may be able to stretch some of the links to suit our short-term needs, and survive the breaking of links as habitats are changed and species are lost, but we cannot escape the fundamental facts of our biological existence: for example, our reliance on photosynthesis, the energy pyramid, and the quality of our physical environment.

The poet John Donne wrote:

> Any man's death diminishes me, because I am involved in mankind.

> And therefore never send to know for whom the bell tolls. It tolls for thee.

We are involved with the natural world, and we are diminished as the natural world is diminished. It is ironic that our ingenuity, our ability to investigate and manipulate the world around us, means that we are destroying some of the beauty of it at the point at which we are beginning to appreciate the true extent of that beauty for the first time. We are a part of Darwin's 'entangled bank', one link in a meshwork of living things, 'so different from each other, and dependent on each other in so complex a manner'.

Study checklist

You should now understand that:

- Populations of plants and animals are controlled by competition for resources, and these resources determine the carrying capacity of an ecosystem for a particular species.
- The population of human beings continues to grow rapidly, as our ability to manage ecosystems allows us to exploit their resources more effectively.
- Human population growth, and our use of ecosystem resources, has had consequences for the biodiversity of ecosystems and the biosphere.

You should now be able to:

- read information from graphs
- appreciate the mixture of scientific evidence and value judgements involved in making decisions about controversial issues.

6 Resources and environmental impact

6.1 Introduction

Chapter 4 showed how we have altered the landscape for our own purposes. People need to cultivate the land, and use its resources, in order to provide food and shelter. In doing so, we have altered the landscape, changing ecosystems to favour the species that are useful to us, such as wheat, cows and chickens. We have used the land in other ways – to build cities, roads and railways, for instance. That can affect populations of other plants and animals, as you saw in Chapter 5. As human civilisation has developed, technology has advanced and so has our ability to alter our environment. For now, I'll just consider the local effects within the UK. Later, in Chapter 7, I'll put this together with the global population figures from Chapter 5, to give a much wider view.

In this chapter, I will revisit the idea of ecological health that you encountered in Chapter 2. I will consider this in more detail, to explore some of the environmental implications of current lifestyles. The main question is: what impact are we having on the environment, and how can we measure it?

Suppose that we go back to Neolithic times (6000 years ago), when people in what is now Britain started farming. A typical Neolithic settlement might have had six families (a total of 30 people). How much land would these people need to support themselves? Food would be a major concern – mainly from wheat, cattle, sheep and goats. They needed timber for fuel, and for building wooden-framed houses. Figure 40 shows the overall land use for this settlement.

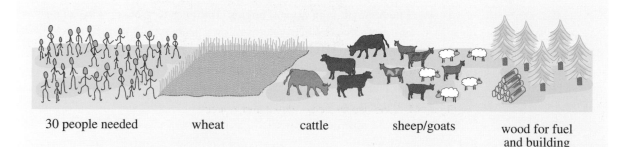

30 people needed   wheat   cattle   sheep/goats   wood for fuel and building

Figure 40 Land use of a Neolithic settlement

A typical settlement of 30 people would have needed:

130 square metres of wheat field

40 cattle

40 sheep/goats

forest area for timber

giving a total area of 6 000 000 square metres of land, which is 200 000 square metres per person (Williams, 2001).

That probably seems like a large area, but the wheat would not have been very productive compared with today's varieties (ancient types of wheat were more like grass), and a large area of forest would be needed to provide a constant supply of wood for fuel.

How does this compare with the land area per person in modern times? This is a complicated question to answer, because most of us no longer grow our own food. We live in cities, and rely on others to provide a steady supply of food, often highly processed, or even pre-cooked for us to reheat. We have other needs: our houses are built from a range of advanced materials – bricks, metal, plastics and glass, as well as timber. In the UK, we expect to have a constant supply of drinking water, electricity and probably also gas to our homes. Our homes, shops and workplaces are heated and cooled to our requirements. We travel by car, train and aeroplane, and many of the goods we use have also travelled many miles. Figure 41 illustrates this.

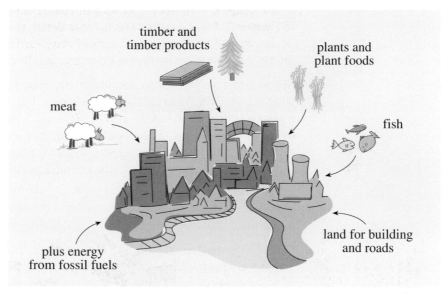

Figure 41   Land use by a city – direct and indirect

Adding all this together is a tricky task. As ever, the approach is to break the complicated problem down into simpler sections. Before I explain how this works, there is one overall point to make: the area used directly (or indirectly) per person for all the activities of daily life is a measure of the impact of that lifestyle on the environment.

One way to think of this is to imagine a dome covering the settlement (or city). Suppose this dome were just large enough to cover the area needed to support the inhabitants of that settlement. For the Neolithic families, the dome would cover an area of 6 million square metres. The question I posed earlier now becomes: how big an area is needed for a typical UK city? We would expect the dome for modern city to be far bigger than for a Stone Age village of 30 people, but how much bigger?

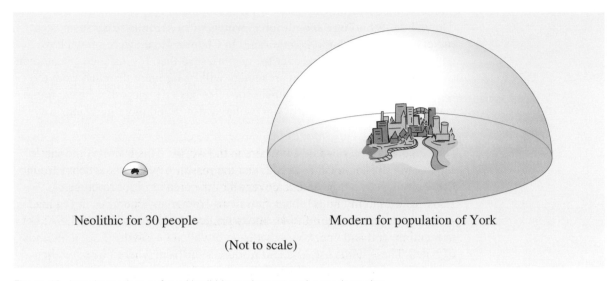

Neolithic for 30 people   Modern for population of York

(Not to scale)

Figure 42   Imaginary domes for a Neolithic settlement and a modern city

6.2 Ecological footprint

The idea of the area needed to support each person has developed into a concept known as the **ecological footprint**, or **ecofootprint**. This is an important idea in discussing the impact a particular lifestyle has on the environment, so I shall be considering it in more detail in this chapter and the next. There is a great deal of information in the short definition below. Read it for yourself, and then I'll explain some of the implications in more detail.

> The ecological footprint is a measure of how much biologically productive land and water area an individual, a city, a country, a region or humanity uses to produce the resources it consumes and to absorb the waste it generates, using prevailing technology and resource management schemes. The land and water can be anywhere in the world.
>
> (WWF, 2005)

The first point is that the ecological footprint is an area – which is why the figures in the previous section were quoted in square metres. I will use other units of area later on, which will be explained as you encounter them.

The phrase '**biologically productive**' may not be familiar, but you have learned about producers and consumers in Chapter 3, so you might guess that this relates to the amount of plant or animal material produced within an area of land or water, for example wheat or fish.

The ecological footprint can be worked out for a whole city, or for each person in that city, which is what I'll cover in this chapter. Similarly, it can be applied to a whole country, or the whole world – the subject of the next chapter.

'Prevailing technology and resource management schemes' may seem rather a dry phrase, but it is important. In Chapter 4, you were shown how technologies have changed over the centuries, so that, for instance, the amount of wheat produced by a given area today will be far more than that from the same area 100 years ago. As human societies have developed, we have found ways to work more efficiently, but we have also found more ingenious ways to use up resources.

'The land and water can be anywhere in the world.' This is one of the subtle aspects of the ecological footprint, and the reason why an actual dome around a real city would not be able to cover all of the area used for resources. Looking around my house, I can find items from many countries: a pen made in Japan; clothes made in China; books printed in Germany and the USA; not to mention fruit and vegetables that have travelled hundreds, if not thousands, of miles. These items use land and resources far from where I live, yet they need to be added to the total impact my lifestyle is having on the environment.

Fortunately, the ecofootprint model has been developed to take account of this complexity. The next section explains how this works.

6.3 Contributions to the ecofootprint

The essential idea behind the ecofootprint is actually quite simple. The area required for each product or process is added up to give the total area, much as you might add up a shopping list. The complicated aspect is that accounting for each item requires detailed research, because it depends on where the item has come from and how it was produced. Each area assumes sustainable production.

As an example of a product, think of one kilogram of potatoes. From the results of research (Barrett *et al*., 2002), I've looked up how much land is needed to grow this quantity of potatoes: see Figure 43. Different researchers may give slightly different results, but this is typical for the UK.

Figure 43 Land area for 1 kilogram of potatoes (each whole square represents 1 square metre)

Source: derived from Barrett et al., 2002, Table 5.3

This may seem a very basic example – that growing 1 kilogram of potatoes requires 2.2 square metres of land – but it serves as a useful comparison. For instance, 1 kilogram of fish requires on average 40 square metres of fishing area (sea, lake or river). That's a far larger area – if it was a surprise to you, think back to Chapter 3, which showed that the further up the food pyramid, the smaller the total amount of biomass produced. A potato plant is at the lowest level of the pyramid (as a plant, it is a producer), so there's more potato per unit area. Fish are consumers, higher up the food pyramid, so there will be fewer fish per unit area, and thus a larger area per fish (see Figure 44).

Figure 44   Water area for 1 kilogram of fish (each square represents 1 square metre)

Source: Derived from Barrett et al., 2002, Table 5.3

What about processed foods? To keep this simple, I'll use butter as an example, because that is made directly from milk, so the main area contribution will be the grazing land for the cows. Typically, 1 kilogram of butter will require 115 square metres of grazing land. It takes a large amount of milk to produce a kilogram of butter, and each kilogram of milk (which would be about a litre) requires 12 square metres of grazing land to feed the cows (see Figure 45).

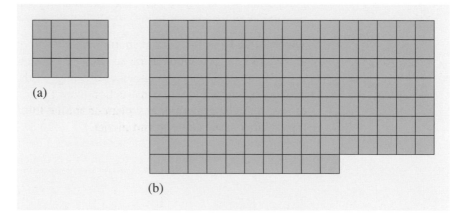

Figure 45   Land areas for: (a) 1 kilogram of milk (b) 1 kilogram of butter (each square represents 1 square metre)

Source: derived from Barrett et al., 2002, Table 5.3

Potatoes, fish and butter are example of food products. To give another example, 1 kilogram of chocolate has a contribution from both grazing land (22.6 square metres) and arable land (41.8 square metres). It contains cocoa and sugar, which both come from plant crops, so these are included as arable land. Most chocolate also contains milk, so grazing land for cows is added.

So far, you have encountered three types of area within the ecofootprint model:

Arable land (sometimes called **cropland**) is used to grow plant crops, such as potatoes and wheat. Orchards would also be included in this category. This is the most productive land type, because plants are at the lowest level in the food pyramid. Plant products are not just used for human food – some are used for animal feed. (Even if the cows producing milk are kept indoors on an intensive farm, the feed would still need to be grown.) Plants also provide biological oils, such as sunflower oil, and some oils and alcohols that can be used as fuel for cars (**biofuel**). Plants supply important non-food fibres, such as cotton and linen, that can be used for making cloth.

Grazing land (sometimes known as **pasture**) supports animals which provide meat, milk, wool, leather and other products. This land is less productive than arable land, because, as consumers, animals are further up the food pyramid.

Fishing area includes freshwater lakes, rivers and saltwater sea fishing. As two-thirds of the Earth's surface is covered by oceans, you might expect an almost unlimited supply of fish. Actually, only a small fraction of ocean area is useful. Most of the fish we eat come from the relatively shallow waters within a few miles of land (the continental shelf), where the fish can find food.

There is another type of land that you came across in the description of the Neolithic settlement: forest. This is the fourth contribution to the ecofootprint.

Forest, which is any area in which trees dominate, can provide timber for construction, furniture and other uses. Wood is still used as a fuel in many parts of the world. Wood pulp, used for making paper, is another important product from forests.

There is one more direct contribution to the ecofootprint, which has become increasingly important over the last 200 years as more people live in cities and urban areas. **Built-up land** includes housing, roads, railways, industrial areas and any other land where construction work has effectively covered the land, diminishing the growth of any plant or animal life. This is the least productive type of area in the ecofootprint model.

Activity 40

Allow about 5 minutes   Types of land in the ecofootprint

What types of land would need to be included in the ecofootprint contribution for each of these items?

(a) Carrots

(b) Wool

(c) Paper

(d) A new road over grassland

Comment

See the comment at the back of the book.

To complete the picture, there is one other important factor to include, but one that is less obvious: energy from fuels. This contribution is more complicated to explain, and to understand it, you need to know more about a natural process called the carbon cycle. Once I've explained that, I'll return to the ecofootprint model, and explain how to add in the contributions from energy use.

6.4 Chemical recycling and the carbon cycle

Chapter 3 looked at ecosystems in terms of some obvious features (looking at the different plants and animals found in a hedgerow), but more especially at what underpins events on the surface – the transfer of energy from organism to organism and from one trophic level to another. You've learned about the movement of energy but also about transformation – for example, solar energy being converted to chemical energy in the form of sugar.

This chemical energy is passed up from one trophic level to the next, as food for the consumers in the next level. At each level, some of the chemical energy is transformed into a form that animals can't use, so it is effectively lost from the food chain. In this type of energy transformation, some of the energy is converted to heat, which is less useful than stored chemical energy. It's just as well therefore that energy in the form of sunlight is available in a (virtually) inexhaustible supply, which constantly replenishes the lowest trophic level – the plants.

In contrast, the chemicals that are essential to life on Earth – carbon, nitrogen and oxygen, for example – are available in fixed amounts. This means that the chemicals such as these that are needed to build and support plants and animals have to be re-used over and over again.

It's an intriguing thought that a tiny amount of the oxygen you are currently breathing in may have been released by photosynthesis from an oak tree in a mediaeval forest a few hundred years ago. Perhaps one of the countless carbon atoms within the carbon dioxide that you've just breathed out was many millions of years ago deep inside the body of a long-extinct dinosaur. Even more remarkable is the likelihood that the very same carbon atom was originally created in what is now a distant star soon after the origin of the universe. (And in this sense humans are no different; we consist of re-usable chemicals of this type.)

This is recycling of materials on such a mind-boggling scale that it is worth looking at in more detail. Once we bring the issues down to earth, so to speak, it will provide another way of expressing the interrelationships between the organisms within an ecosystem.

Let's start with a 'mini' carbon cycle – the organic carbon that flows between animals and plants. You encountered this in Chapter 3, when producers and consumers were discussed. Figure 46(a) shows the movements of gas between animals and plants.

Look at Figure 46(b). You can see that it's a reworking of Figure 46(a), with
more detail and clarity. Note that each box represents a store, or **reservoir**, of
a form of carbon, and the arrows represent processes which transfer carbon
between the reservoirs. In the atmosphere, carbon exists as carbon dioxide gas,
which is involved in photosynthesis and respiration. What also links the 'plant'
and 'animal' boxes is carbon in the form of complex chemicals, eaten as food
by the consumers (animals). Although I won't do so here, it would be possible
to estimate the amount of carbon within any of the boxes or the rate at which
carbon is added to or removed from a particular box. This could be either on a
local scale – within a particular ecosystem – or on a grander, global scale.

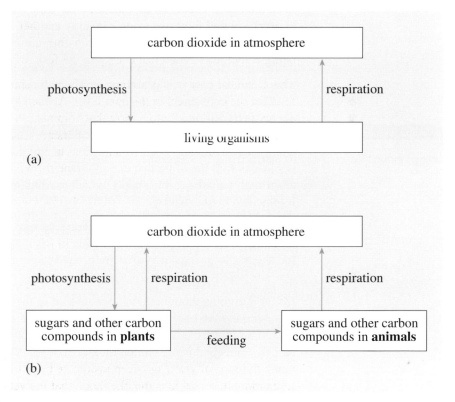

Figure 46   (a) Exchange
of gases between plants
and animals (b) A simple
model of the carbon cycle

Let's now add to this simple picture step by step.

- You'll appreciate from Figure 46(b) that a good deal of carbon is 'in
  transit' between various storage sites – that's what the arrows represent.
  But at any one time, far more carbon is locked up in such stores than is in
  circulation. And there are more stores than are shown in Figure 46.

- In particular, a great deal of carbon is locked up in rocks – much of it
  within the calcium carbonate that is the main constituent of limestone
  and chalk. Exposed rocks of this type are perpetually broken down (or
  weathered), and this releases carbon dioxide into the atmosphere.

- Roughly the same amount of carbon is deposited as chemicals in
  sediments in the oceans – much of this comes from marine animals'
  shells. Gradually, over many years, this sediment is transformed into

rocks, such as chalk. A great deal of carbon dioxide is held in storage in the ocean where the gas dissolves in sea water. Some of this carbon slowly returns to the atmosphere.

• Volcanoes release some carbon from rocks into the atmosphere.

• Once living organisms die, much of their carbon returns to the soil. Vast amounts of carbon of animal and plant origin are contained in soil. Some of this carbon can be released into the atmosphere as carbon dioxide, although much of it remains in the soil. Soil can also be gradually transformed into rock. As a result of this, some rocks also contain the remains of organisms from the geological past, and these have produced fossil fuels – coal, oil and natural gas.

The whole cycle is more or less in balance, with the total amount of carbon going into each carbon store roughly equal to the total amount leaving. This applies to all the components of the cycle – living things, the atmosphere, the oceans and the land (rocks and soils). The natural carbon cycle is balanced.

Activity 41

Allow about 15 minutes   Producing a more detailed carbon cycle

Copy out Figure 46(b) and adapt it to show the flow of carbon involved with the processes in the list above. For simplicity, include plants and animals within a single box, labelled 'living organisms', to the left of your diagram. Include another labelled 'oceans' on the right. 'Rocks, fossil fuels and sediments' can all go into the same box, at the base of the diagram. You will also need a box labelled 'soil'.

Comment

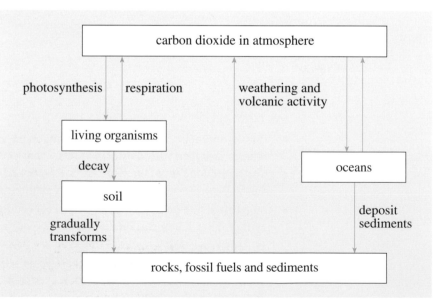

Figure 47   A more detailed model of the carbon cycle

Your diagram no doubt differs from Figure 47, but I'm more concerned about whether you found the exercise useful. What these diagrams emphasise are the interrelationships between the various components of the system – something of special relevance here.

Compare the description of the carbon cycle in words with Figure 47. Which do you find easier to remember?

With flow diagrams that form a cycle, it's possible to start at any one box (say that for atmospheric carbon dioxide) and then remind yourself of all the main points by working round step-by-step in one direction, until you're back at the beginning. Try this for the carbon cycle, without reference to the previous text. How much can you remember?

Again, the important point is that diagrams are useful tools to help you learn. For example, working your way through Figure 47, referring to the text where necessary, should help to reinforce your understanding of the key ideas covered in this section.

Upsetting the balance – fossil fuels

If, as I've implied already, the carbon cycle is balanced, the amount of any one component is likely to be roughly constant. So, for example, you'd expect the amount of carbon dioxide in the atmosphere to be constant, at least over a short period of time. Of course, it isn't the same carbon that is always present as atmospheric carbon dioxide; it's more a case that if the rates of arrival of 'new' carbon dioxide (from respiration, weathering and release from the ocean) are equal to the amounts of leaving for other sites (via photosynthesis and dissolved in the ocean) then, over time, the amount of atmospheric carbon dioxide should remain about the same.

Fossil fuels

Fossil fuels contain a large amount of carbon that is released as carbon dioxide gas whenever these fuels are burned. Coal is almost solid carbon, although it contains other elements, such as sulphur, that can form poisonous gases. Petroleum oil and gas ('natural gas') are cleaner than coal – they produce fewer polluting residues such as sulphur when burned – and they are easier to purify because they are fluids (liquids and gases).

Most fossil fuels were formed from plants that lived at a particular stage in the Earth's history – the Carboniferous period (290–354 million years ago). At that time, there was rich vegetation over large areas, which then became covered in mud and other deposits. These deposits buried the decaying plants, preventing them from decomposing completely. Over millions of years, the plant deposits were buried deep underground and then subjected to heat and pressure, forming coal, oil and gas. It could be said that fossil fuels are 'stored sunlight', as the chemical energy they contain was originally converted from sunlight by the plant life of the Carboniferous period. (See Appendix 1 for a geological time scale.)

Fossil fuels are no longer being actively formed on a large scale – the natural processes that made them depend on particular climate conditions, and take millions of years. Although there are disagreements about the amount of fossil fuel left, it is clear that it is a unique supply of fuel for the planet – once used up, it will not be replaced. Hence, fossil fuels are often referred to as non-renewable resources. A fuel reserve is the amount of that resource that is economically recoverable. Even if there is a large amount of oil in a given area, it may not be technologically possible to recover it, or doing so may simply be too expensive.

Oil is particularly useful because it contains a large amount of energy per litre, and it is easily transported and processed (to make petrol, for instance). In a car's engine, the chemical energy stored in the fuel is converted to motion (mechanical energy) and heat. Imagine carrying one tonne of metal over a distance of 15 miles in 20 minutes. Now consider that a typical car weighs about a tonne, and can do exactly that, using the energy from less than 1 litre of petrol.

Oil is also used as the raw material for plastics and many other carbon-based chemicals, such as medicines, polyester fibre for clothing, paints, plastics, glues and printing inks. This book could not have been made without oil.

The implications of burning fossil fuels for the levels of atmospheric carbon dioxide are now well known: pollutants are being released into the atmosphere and carbon dioxide is a greenhouse gas that may be increasing global warming (discussed later in this chapter). Less well known are the effects of clearing tropical forests. The burning of the cleared vegetation converts most of the biomass to carbon dioxide. If the cleared area is subsequently ploughed for agricultural use, carbon dioxide loss is much more substantial because the decomposition that results releases much of the carbon that was previously locked up in the soil. The amount of carbon involved makes a significant difference to the carbon cycle. Each year, the burning of fossil fuels releases about 20 times as much carbon into the atmosphere as the natural processes of weathering and volcanic activity combined. The clearing and burning of forests releases up to half as much again, although this figure is difficult to determine accurately. The actual amounts of carbon involved are huge: each year, volcanoes release about 50 million tonnes of carbon into the atmosphere, weathering releases another 200 million tonnes and burning fossil fuels releases about 5400 million tonnes.

Allow about 40 minutes

Environmental shopping

Carbon dioxide emissions

Track 4 of the DVD is an interactive video activity about carbon dioxide emissions. Using a trip to the supermarket as an example, you are asked to make a series of three decisions that will affect the use of fuel, and the emissions, for the trip:

1   the type of transport to the supermarket: car, bus or walk

2   buying local or imported food

3   carrying the food home in a disposable bag, a reusable plastic bag or a cloth bag.

Follow the instructions on screen, using the DVD menu controls to make your choice at each stage. At the end, the DVD will automatically select a summary, depending on whether your choices lead to high, medium or low emissions. There are many factors that can affect emissions – the examples shown are intended to illustrate the types of issues that are involved, and to show you some of the wider implications. The information included in the video is from a variety of sources, and is intended to be indicative for the UK.

You may like to simply play with the options, to see what happens. Make a note of your choices, and the information presented in the video clips, as you go through. What other factors affect people's choices?

Also, find a pathway (there are several possibilities) that leads to:

(a)   low emissions

(b)   high emissions.

Comment

There is no 'right' answer for this activity, although there are suggestions for ways in which carbon dioxide emissions could be reduced. As you have discovered, our everyday decisions are affected by many factors other than simply reducing the amount of fuel used. There are also scaling factors – if many people use a bus rather than individual cars, it adds up to an overall saving. In contrast, if many people choose to use disposable packaging, even though the amount per person is small, it adds up to a large increase in emissions. Note that the results of the video activity are just indications – they are not based on detailed calculations.

What happens to all the extra carbon dioxide in the atmosphere? About one-fifth of it dissolves in the oceans, which form a reservoir, a store of carbon. Another two-fifths leaves the atmosphere in other ways, probably taken up by extra tree growth. But so much carbon dioxide is currently being released as a result of human activity that it appears that the capacity of the oceans and trees to mop up excess carbon dioxide has been exceeded – the gas now accumulates in the atmosphere. Figure 48 (overleaf) shows what is happening.

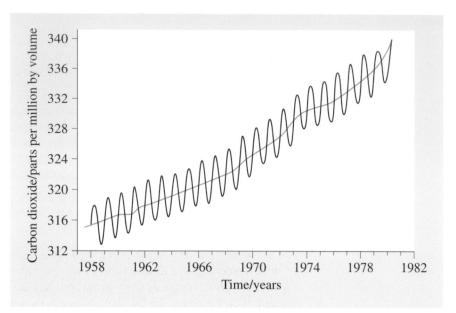

Figure 48   Levels of atmospheric carbon dioxide showing seasonal variation, recorded in the atmosphere of Hawaii

Source: Keeling and Whorf, 2005

Activity 43

Allow about 15 minutes

Dealing with graphs

What 'sense' can you make of Figure 48? When you need to make sense of an unfamiliar graph, you should start by reading the title and the labels on the two axes. Always look for the simplest message the graph conveys – the overall shape of the line or curve. Is it increasing or decreasing? Does that line stay flat, or is there some variation?

Write down in words what you can about the relationship shown in the graph.

Comment

This graph shows the variation in carbon dioxide levels in the atmosphere over time, between the years 1958 and 1980. The overall shape slopes upwards to the right, showing that carbon dioxide levels in the atmosphere have gone up. Between the years 1958 and 1980, the amount of carbon dioxide increased from about 315 to about 340 parts per million by volume. That's an increase of 25 parts per million by volume.

Measuring very small dilutions

One part per million by volume of carbon dioxide means that for every million units of gas, just one is carbon dioxide. 'Parts per million' is sometimes written 'p.p.m.'.

This is difficult to imagine, but one way to get a sense of it is to consider a simpler example. Imagine a domestic bathtub full to the brim with water (that's about 200 litres, or one-fifth of a million millilitres). A drop of ink added to this would be diluted to one part in a million (assuming the ink drop had a volume of one fifth of a millilitre).

So one part per million is equivalent to a drop of ink in a full bathtub – a tiny amount.

The seasonal 'ups' and 'downs' in Figure 48 are due to changes in plant growth. Remember the descriptions of photosynthesis and respiration from Chapter 3 – photosynthesis uses up carbon dioxide. In the summer, the greater amount of carbon taken up by plants reduces the carbon dioxide concentration in the atmosphere. In winter, there is less plant growth; this, together with continued release of carbon dioxide from respiration, leads to increased atmospheric carbon dioxide.

But what of the overall trend? In 1958, carbon dioxide concentration was about 315 parts per million by volume (see the box above). This means that in each million cubic metres of air, there were 315 cubic metres of carbon dioxide. In 1990, the concentration was close to 350 p.p.m., and back then the prediction was somewhere between 375 and 400 p.p.m. for the end of the century (the year 2000). With the benefit of up-to-date information, we can see whether this scientific prediction was correct. Figure 49 shows the latest results from the same measuring observatory in Hawaii.

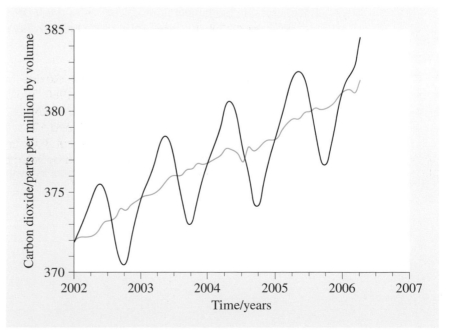

Figure 49   Current results from Hawaii

Source: Keeling and Whorf, 2005

Allow about 5 minutes

Testing a prediction

From Figure 49, what is the current level of carbon dioxide in the atmosphere?

Was the prediction made in 1990 correct?

Comment

See the response at the back of the book.

The greenhouse effect and climate

As you've seen, these are, overall, very dilute amounts of carbon dioxide in a large volume of atmosphere. So, do these changes make any difference?

You may know that carbon dioxide and other gases in the atmosphere trap the sun's energy and keep the Earth's surface warm: this is known as the greenhouse effect. Were it not for these 'greenhouse gases', heat would escape back into space and the Earth would be 33 °C cooler than it actually is, which would mean that much of the Earth's surface would be deeply frozen. What is less certain is whether the higher carbon dioxide levels have resulted in an extra warming of the Earth's surface – so-called global warming. Releasing carbon that has been stored away in coal, oil and gas is bound to affect the carbon cycle in some way, especially as this has happened over the relatively short time scale of a couple of hundred years. The question is – how much difference is this making? Some critics argue that other factors explain why we've had so many 'warm years' in the past decade or so. For example, changes in the amount of solar energy hitting the Earth's surface may be increasing the temperature. These critics don't deny the evidence that carbon dioxide levels have changed, but they believe that such changes have little to do with global temperature.

But these reservations are the exception. Although the extra carbon dioxide has been diluted by the atmosphere, these changes do appear to be making a difference to the world's climate. There is a strong consensus among scientists that despite a number of uncertainties, global warming is a present-day reality and will be an environmental threat of increasing severity to the next generations, unless action is taken to reduce the rates of release of carbon dioxide and other greenhouse gases into the atmosphere.

Oceans and climate

Over two-thirds of the Earth's surface is covered by oceans, which play an important part in the world's climate. Carbon dioxide and other gases dissolve in water, so the oceans are a vital part of the carbon cycle. Water has some distinct properties that are essential for life on Earth. Unlike many materials, solid water (ice) floats above the liquid, so the frozen polar regions have vast white sheets of floating ice and icebergs. These help to reflect the sun's heat, keeping those regions cooler. In regions where lakes freeze over in winter, life can continue in the water beneath.

Water also acts as a very effective heat store – a comparatively large amount of thermal (heat) energy is needed to increase the temperature. So, the oceans effectively even out the temperature variations between the seasons in coastal areas. Further inland, the temperature variations are far more marked. On a global scale, ocean currents circulate warm water from the equator to the cooler areas near the poles. For example, the Gulf Stream that runs northwards across the Atlantic, up the west coast of Britain, is responsible for keeping the UK's climate relatively mild. One of the concerns, and debates, among climate scientists, is whether this '**thermal current conveyor belt**' could be affected by human activities, possibly causing climate disruption.

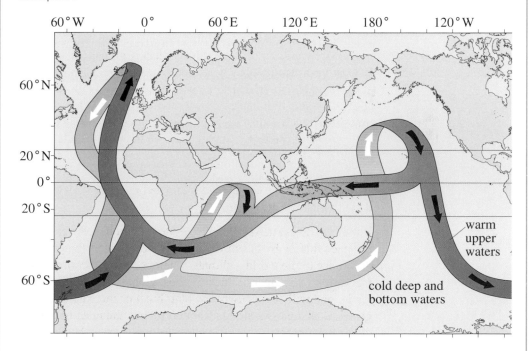

Figure 50   The thermal current conveyor belt – ocean currents that circulate warm and cold water around the globe

6.5 Adding fossil fuels to the ecofootprint model

Any process that uses fossil fuels will add carbon dioxide to the atmosphere. So, the impact of this contribution needs to be added to the ecofootprint model. Section 6.3 introduced five types of area that are used in this model: arable land, grazing land, forest, fishing area and built-up land. The fossil fuel contribution is added as an amount of carbon dioxide released, but that is then converted to an equivalent area of land.

What matters is that according to this model the more carbon dioxide released into the atmosphere, the worse the environmental impact. The energy contribution to the ecofootprint differs from the other area types, because it adds up the carbon dioxide emissions from fossil fuels. The model assumes that the oceans absorb a significant amount of carbon dioxide, so the 'energy contribution' is the amount required to make up the difference. This can make a significant contribution to the overall ecofootprint, particularly in countries where there is a high use of fossil fuels.

Although the physical effect is the release of carbon dioxide into the atmosphere, the model needs to include this as a 'land' type, so that it can be added on to the other contributions for grazing land and so on. This may seem odd, but it is just a way to include the effects of carbon release. So, you will sometimes encounter the fossil fuel energy contribution as 'tonnes of carbon dioxide', or as an area of 'energy land', depending on how the model has been applied. To convert from one to the other, the model assumes a theoretical area of vegetation that is needed to absorb that much carbon dioxide. (Remember the lower carbon dioxide levels during the summer, in the graphs from Hawaii.) 'Energy land' does not actually exist – it is a theoretical area, showing how much new forest would need to be planted to neutralise the effects of the extra carbon dioxide released.

For example, 1 kilogram of butter is processed from 12 litres of milk, using some fuel. As most industrial processes in the UK are based on fossil fuels (much of our electricity is generated from oil, coal or gas in power stations), this use of fuel will release some carbon dioxide. As for the other contributions to the ecofootprint, there are tables of results available, showing how much carbon dioxide is released for each process. For 1 kilogram of butter, the extra 'energy land' to absorb the carbon dioxide produced by the use of fuel adds another 59 square metres to the ecofootprint, in addition to the 115 square metres of land required to produce the milk (Barrett et al., 2002, Table 5.3).

Any process that uses energy from burning or using carbon-based material will add a component of 'energy land' to the ecofootprint. This includes fossil fuels as well as wood, cardboard or paper. It also takes account of mains electricity, if it has been generated from burning coal or gas. Nuclear, wind, hydroelectric and other alternative sources do not contribute to carbon dioxide emissions. Despite this, in some ecofootprint models, including the one from the WWF quoted in this book, nuclear power is added in as a contribution to the ecofootprint because of the environmental effects of radioactive waste disposal.

So, the carbon dioxide amounts can be shown as their (theoretical) land equivalents. This 'energy land' can then be added onto the overall ecofootprint. Figure 51 shows Figures 43, 44 and 45 redrawn to include this theoretical energy land (shown as pale grey squares).

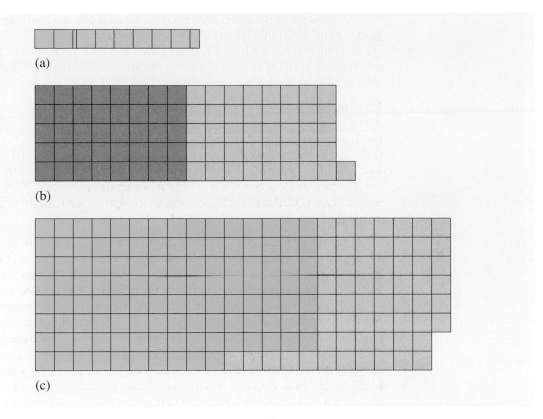

Figure 51   Total ecofootprint, including 'energy land' for 1 kilogram of (a) potatoes (b) fish (c) butter (each small square represents 1 square metre)

Source: Barrett et al., 2002, Table 5.3

Activity 45

Allow about 15 minutes   Energy land

From Figure 51, compare the relative amounts of energy land for the different products, and suggest why these values differ. Think about what is involved in growing, harvesting, storing and processing the products, and how much fuel might be needed.

Comment

Potatoes: these need the least processing, and are stored at room temperature, but include some fuel for farm machinery.

Fish: not highly processed, but needs to be chilled or frozen, and also requires fuel for fishing boats.

Butter: higher use of fuel, because there is more processing from milk, and must be chilled.

'Energy land' is rather an abstract concept, but it is useful for comparing different processes, as you will see in Chapter 7.

6.6 Using numbers and units

We all use units of measurement in our everyday lives – kilograms, litres, seconds and so on. There are hundreds of different measures, but scientists and technologists of all nations have agreed to use a standard system of units. Everything you could ever want to measure can be measured using a few basic units, or combinations of them. These internationally agreed units, and the rules for their use, form the Système Internationale (international system), known as SI. This section explains some of the essentials of SI, which is sometimes referred to as the metric system. If you continue your studies, particularly in science or technology, will encounter more examples of the use of SI units. In some situations, however, you may also encounter non-SI units; for example, in the UK, miles are often used to measure longer distances.

To show you how the SI system works, let's look at the units of length. The SI base unit for length is the metre, symbol 'm'. It is useful to have a rough idea of the size of a typical measurement in any given unit, so, for example, the height of a standard domestic door is about two metres. If you are not familiar with the SI system, get some practical experience of the size of different units by noting where they are used in your life. For longer lengths, such as the distance between two towns, the SI system uses a multiple of the base unit, a kilometre (symbol 'km'). The 'kilo' in kilometre is an example of a prefix – a letter or word placed in front of another word or symbol. Prefixes show the scale of the measurement, and are standard for all SI units; 'kilo' means 'a thousand'. So one kilometre is a thousand metres.

There are other prefixes for smaller units. So, the length of this page might be measured in centimetres (cm), where 'centi' means 'a hundredth'. A centimetre is one-hundredth of a metre, so there are 100 centimetres in metre. A UK one penny coin measures about 2 cm across. For even smaller distances, we use millimetres (mm). The prefix 'milli' means 'a thousandth' so a millimetre is one-thousandth of a metre (or a tenth of a centimetre).

Units of area are particularly useful in this book. These are illustrated in Figure 52. The SI unit of area is the square metre (symbol 'm²'), although multiples of this are often used for larger areas. A square kilometre (1 km²) is an area one kilometre by one kilometre, i.e. 1000 metres by 1000 metres. There are 1 000 000 square metres in one square kilometre. The ecofootprint and other land areas are often measured in hectares (symbol 'ha'). One hectare is an area 100 metres by 100 metres, so there are 10 000 square metres in a hectare and 100 hectares in a square kilometre. Appendix 2 shows the equivalents for various units.

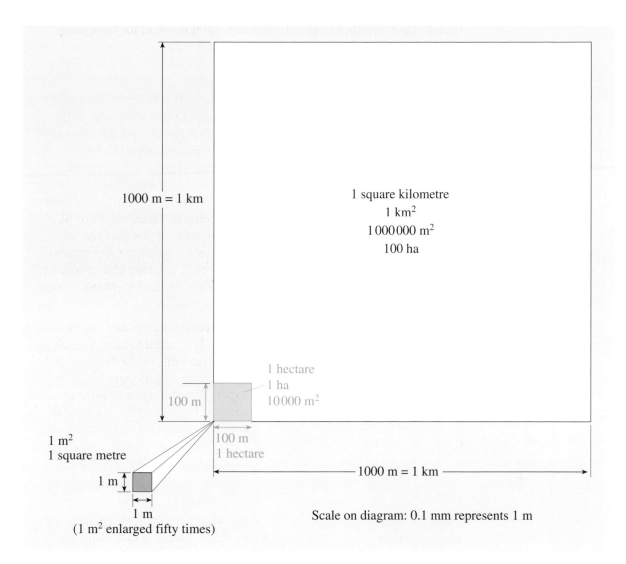

Figure 52   Comparing units of area

Allow about 20 minutes   Converting units

Using the information in this section and Appendix 2, convert the following units:

(a) Write five hectares as square metres

(b) Write eight square kilometres as hectares

(c) Write 230 square kilometres as square metres

Comment
See the response at the back of the book.

So far in this section, I have only considered measurements of length and area. The full SI system has base units for many other quantities, such as mass and time. The common unit for mass is the **kilogram** (kg), although grams (g) are often used. The prefix 'kilo' has the same meaning, so one kilogram is 1000 grams. Sugar is often packed in 1 kg bags. For larger masses, such as the mass of a car, the **metric tonne** (symbol 't') is used, where one tonne is 1000 kilograms. Note the spelling – if you see 'tonne' written by itself, that means metric tonne. If you see 'ton', that implies the older, non-metric ton. Strictly, tonnes are metric but not SI, although they are often used alongside the SI units for convenience. Sometimes you may hear people talking about their 'weight' in kilograms. This is another instance where the everyday use of a word differs from the technical or scientific definition. There is a difference between mass and weight – your **mass** is the amount of matter in you, but your **weight** depends on the gravitational pull on you as well. When you see a measurement called 'weight' and quoted in grams, kilograms or similar units, it should strictly be called 'mass'.

6.7 Ecofootprint as a model

The ecofootprint is an example of a **scientific model**. A model represents some aspects of the world, while omitting others. It is a useful way to simplify situations so that predictions can be made. For example, the schematic map of the London Underground train routes shown in Figure 53(a) is useful, because it shows the connections between the different lines, and the order of the stations. If you look at another model for this area of London – the street map shown in Figure 53(b) – you see different information. Here, the exact location of each station is shown, but it is hard to tell which way the train lines go. So these two models each have advantages and disadvantages, but they are both useful, depending on the circumstances, and neither is 'wrong'.

Summarising the relevant features of a situation creates a model – a simplified version of the system – that we can use to look at the problems or questions in that situation. Some models are physical (a toy train, for example), others may use diagrams (a map), while others use numbers or ideas to represent the world.

Whenever you consider a model, remember that it is a representation of reality, not the actual world. There are three main issues to consider:

- what the model is used for
- the ways in which the model matches reality
- the limitations of the model.

A scientific model should be a good match to the actual situation, because it will have been tested using rigorous methods. The ecofootprint is generally regarded as a useful model, because it accounts for many of the effects of human activities on the environment, in a way that can be checked and measured. There are different versions of the model, and there is some debate about the details. As more research is carried out, the ecofootprint model will change, to take into account new discoveries.

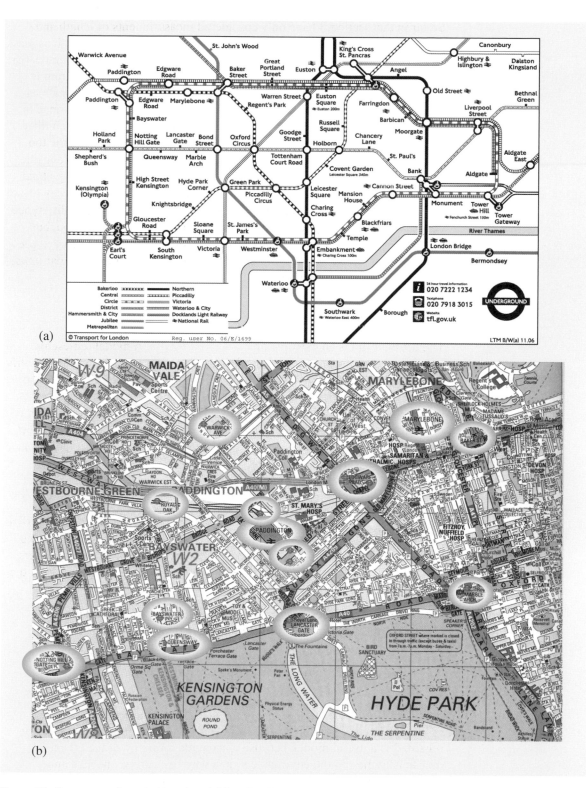

Figure 53   Two maps of central London: (a) London Underground train map (b) street map showing Underground stations

What is not included in the ecological footprint model?

The ecological footprint is based on measures of sustainability. It does not currently include:

- activities for which there is no reliable information, such as acid rain, even if these are known to have an impact on the environment (although such activities could be included in the future if data became available)

- the use of materials that are not part of natural cycles, for example the plutonium produced by some nuclear power stations

- processes that cause irreversible damage to the biosphere, for example, species extinction or fossil fuel depletion

- any space for wilderness, in other words, the assumption is that all available land area is used for human activities, with no nature reserves or wild areas other than 'useless' land, such as deserts. This would not be environmentally desirable, so some ecofootprint models do allow a certain proportion of wilderness.

This means that overall, the ecological footprint underestimates the environmental impact of human activity.

6.8 Current resource use in the UK

Going back to the idea of a dome covering a city that you met earlier in this chapter, how big would the dome be for an actual city, such as York? The council of the city of York carried out a detailed analysis of the ecofootprint in the year 2000 (Barrett et al., 2002). This section uses information from that survey, so it is about an actual UK city, not an imaginary location.

In the year 2000, York had about 179 000 residents, and the total ecofootprint per resident was 6.98 hectares (698 000 square metres). Multiplying the ecofootprint per person by the number of people gives a total ecofootprint area for the city of 1 249 420 hectares (124 942 000 000 square metres). This is far larger than the actual area of the city, the built-up land. The actual area covered by York is 27 200 hectares, so the ecofootprint is much bigger than the area covered by the buildings. This is what you'd expect, because the city will import goods from elsewhere in the UK, Europe and further afield.

Figure 54(a) shows the area contributions to the ecofootprint of York, with built-up land at the centre. The built-up land is the actual area occupied by the buildings and roads of the city. Figure 54(b) adds an extra ring of area, to represent the theoretical 'energy land' due to fossil fuel use in the city – notice how much this adds to the overall ecofootprint. This shows the total environmental impact (and resource use) for York. If it were possible to put a sealed glass dome over the city, it would need to encompass the area shown in Figure 54(b). This imaginary dome is shown in Figure 54(c), over a map of

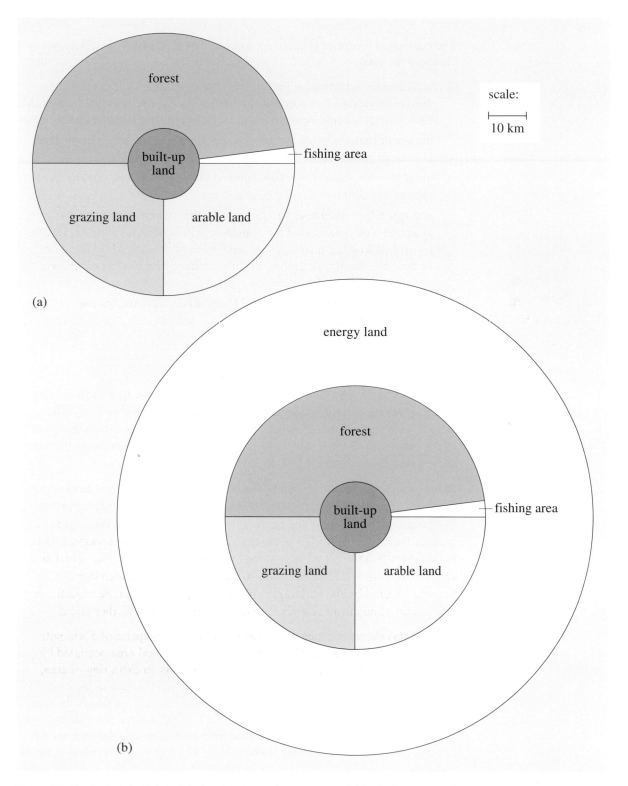

Figure 54   Ecofootprint of York: (a) showing five main area types (b) including 'energy land'

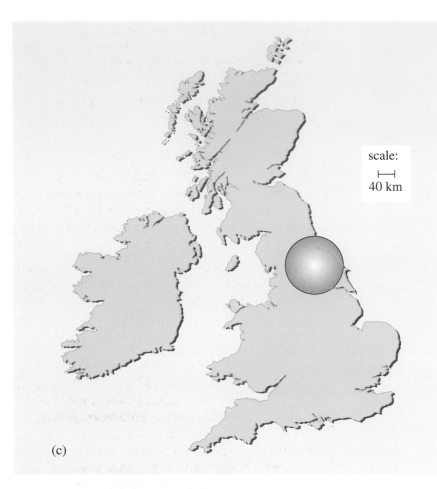

scale:

⊢—⊣
40 km

(c)

Figure 54(c) Ecofootprint of York shown as an imaginary dome over northern England

Britain to the same scale. Of course, this dome is just to illustrate how much resource the city needs. Because food and other resources are imported from all over the world, the actual contributions to the ecofootprint are scattered in different countries. For instance, if vegetables are imported into York from South Africa, their arable land contribution is part of the York ecofootprint, even though the land is actually thousands of miles away.

You have already seen how the ecofootprint is made up from contributions of different types of area. Table 4 shows how this works for York, and for the UK as a whole, so we can compare them. This shows the ecofootprint per person.

Table 4 Ecofootprint of York, and average ecofootprint for the whole UK		
	York ecofootprint /hectares per person	UK ecofootprint /hectares per person
Energy land	4.68	3.79
Grazing land	0.49	0.69
Arable	0.51	1.00
Sea	0.03	0.05
Forest	0.93	0.36
Built land	0.15	0.37
Total	6.98	6.26

Source: based on Barrett et al., 2002

Activity 47

Allow about 5 minutes

Is York typical of the UK?

Compare the figures for York with those for the whole UK. What similarities and differences can you find?

Comment

Comparing each row in the table, the figures are fairly similar, although the overall value for York is higher than for the UK as a whole. Per resident, York uses more energy land and forest, and less arable land, than the UK overall.

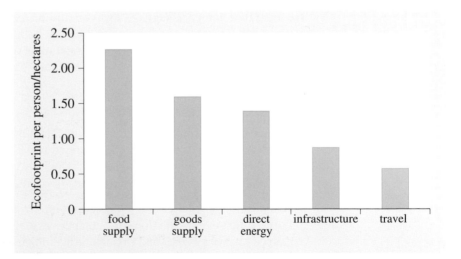

Figure 55 Ecofootprint of York in the year 2001, divided into the main contributing factors

Source: Barrett et al., 2002, p.85

The ecofootprint for a city can be examined in various different ways. In this section, we'll look at some of the contributions to this ecofootprint. Later, in Chapter 7, you will see how changes might be made to reduce the environmental impact of some of these contributions.

Most of the ecofootprint of York was due to food products – in one year, the city used 92 500 tonnes of food and drink. As you saw on DVD Track 4, there are contributions beyond the actual land needed to produce this food. Out of the 2.3 hectares per person for the food supply, only 0.9 hectares represents the actual land area required to grow or rear the food. Another 0.99 hectares of energy land is required to account for the fuel used in farm machinery, food processing and transporting the food. Food packaging accounts for another 0.35 hectares out of the total 2.3 hectares.

The detailed information for 'energy land' provides some more clues about how the residents of York are using resources. Remember that 'energy land' represents the amount of carbon-based fuel used. The total 'energy land' for the city is roughly 864 000 hectares. The complete list of all contributions to this total contains over 150 separate items, from air travel to yoghurt production (Barrett *et al.*, 2002). The interesting point is that out of all of these, just six entries account for over half of the total carbon dioxide emissions. These 'energy land' entries are (to the nearest 1000 hectares):

Domestic gas use	109 000 hectares
Car travel	97 000 hectares
Domestic electricity use	79 000 hectares
Brick production	71 000 hectares
Commercial paper and card production	48 000 hectares
Commercial electricity use	48 000 hectares

So, next time someone suggests that you turn the central heating thermostat down a few degrees, switch off a few lights at home, and use the car less, you'll know why. Even a small proportional reduction in each of these would make a large difference to the overall environmental impact of the city.

The ecofootprint due to travel and transport

Figure 56(b) shows the contributions of different types of travel by the residents of York. Note that this means any travel by people living in the city, not just movement within the city boundaries. Also, it only includes people travelling. Commercial transport of goods is not included here.

Cars have the most impact on the ecofootprint, owing to a combination of a relatively high carbon dioxide emission per passenger mile (which has to be balanced by energy land) and frequent usage. Planes have a slightly higher ecofootprint per passenger mile than cars, and are used for longer journeys, but less frequently than cars. So, one plane journey can make a relatively large contribution, but when averaged out over the year over all inhabitants of York, this is less significant than car use. The car ecofootprint also includes a contribution from built-up land, to account for the areas of roads required – over three-quarters of road vehicles are cars, so that proportion of road area is added to the car ecofootprint. The ecofootprint also takes into account the resources used to make new vehicles and dispose of old ones.

Figure 56(a) The city of York (b) Ecofootprint contributions from travel by York residents in the year 2001

Source: derived from Barrett et al., 2002

Table 5 Comparing carbon dioxide emissions for a journey from London to Edinburgh		
	Amount of carbon dioxide produced per passenger for the journey/kg	'Energy land' equivalent per passenger for the journey/m^2
Plane	96.4	318
Car	71.0	234
High-speed electric train	11.9	39
Coach	9.2	30

Source: *Guardian*, 2005

It is also possible to compare different means of transport, for example, just in terms of their emissions. Sometimes this is expressed directly in terms of mass of carbon dioxide released per mile, but it can also be expressed in terms of 'energy land' required, so that it can be added directly to the ecofootprint. In the ecofootprint model, 0.33 hectares of 'energy land' are needed per 1000 kilograms of carbon dioxide emitted. The more fuel used, the more carbon dioxide emitted. Table 5 shows some typical values for a journey of about 400 miles, so you can see the difference between some common modes of transport.

Activity 48

Allow about 15 minutes

Transport and carbon dioxide emissions

As Table 5 shows, flying produces far more carbon dioxide per passenger than the other types of transport. Planes require more fuel than other vehicles because of their size and speed and just to stay airborne. Compare the values for a car and a coach. Why do you think the coach is more efficient?

Comment

The amount of fuel used per mile for a car and a coach may differ, but the main factor is that the coach carries far more passengers, so the emissions per person are lower. Similarly, four people travelling in one car is about four times as efficient as four cars each with just one person. This is why some organisations promote car sharing as a means to reduce emissions (and save fuel and money). Of course, the coach journey will take far longer than flying from London to Edinburgh. If the time taken to check in (and wait for luggage) at the airport is included, the train journey takes about the same overall time as flying.

This chapter has looked, in some detail, at the ecofootprint model. You have seen how the different contributions can be added up for a kilogram of food, a single journey, or a whole city. Although there are some complicated details, the basic idea is simple – this model allows us to add up the use of resources, and thus estimate the environmental impact. The more resource we use, the greater the demands on the environment, so we should try to make the ecofootprint as small as possible. All models have their benefits and limitations – the advantage of this one is that it gives us a basis for comparing different activities, so useful decisions can be made. In the next chapter, this idea will be extended to whole countries, so you can compare resource use with the resources available, on a global scale.

Study checklist

You should now understand that:

- The ecofootprint is a scientific model – a representation of the world.

- Elements, such as carbon, are recycled in nature, with a delicate balance between different stages of the cycle.

- Human activities, such as deforestation and burning fossil fuels, have altered the balance of the carbon cycle. The detailed results of this are difficult to predict, but global warming is a serious concern.

- The contributions to the ecofootprint of a community (such as a city) can be calculated, to see which activities make the most difference.

You should now be able to:

- relate a diagram to a description in words

- redraw a diagram to include extra information

- describe and interpret the shape of a line graph

- read information from a table

- compare numbers using similar units of measurement

- review ideas you have encountered previously, and relate them to new ideas presented here.

7 Sustainability and our environment

7.1 Introduction

The previous chapter introduced the idea of an ecological footprint, and showed how it works for the city of York. Ecofootprints can be determined at different levels, for a single household, city, region or whole country. In this chapter, the ecofootprint will be extended to the whole of the UK, and compared with other countries in Europe and the rest of the world. The ecofootprint shows how we use resources, and it is important to compare this with the amount of resource actually available. Are we living within our means?

This chapter addresses this question by introducing two new ideas: biocapacity and sustainability. Biocapacity tells us what resources are available. This is about the reality of actual land and water area – we cannot create extra continents or oceans, although we may be able to plant more forests. Sustainability compares what we are using with what is available. You have already encountered 'maximum sustainable yield' (MSY) in Chapter 5, where it was applied to cod stocks. If we use fewer resources than are available, then we are 'in credit' – living at a level where, overall, the environmental impact is low enough for us not to cause permanent damage. This is sustainable living. If, however, we use more resources than are available, then we are 'in debt', so that the damage to the environment will increase, without recovery – this is non-sustainable living. To work out what's sustainable, and what isn't, I'll show you how to compare the ecofootprint (what we're using) with the biocapacity (what's available). This approach can be applied more widely than MSY, because it relates to more than a single species such as cod.

Then, I'll look at some ways to reduce the ecofootprint, to make lifestyles more sustainable. In the previous chapter, you saw how different forms of transport have different environmental impacts, so the ecofootprint can be reduced if people choose to use transport differently. This chapter will consider two further examples. The first of these is housing – Chapter 6 showed that domestic heating and electricity use are a major contribution to the ecofootprint of a typical city. Later in this chapter, with the help of a video documentary, you can look in more detail at ways in which housing and lifestyles affect the environment. The second example is waste disposal. It may seem dull at first, but it can have widespread environmental implications. Earlier chapters have shown how natural cycles of growth, decay and decomposition allow nutrients to be recycled, for instance in a compost heap or on a farm. Our society does not necessarily recycle waste in a way that allows material to be re-used.

7.2 How much useful land and water area is actually available?

Just as Chapter 6 described a measure of environmental impact (ecofootprint), this section describes a measure of useful biological resource (biocapacity). By useful, I mean land or water that supports an active community of plants and animals, is self-sustaining, and can produce the surplus needed for food, timber and other human requirements. Chapter 3 introduced the ideas of biological production and biomass. At that stage, you saw how food chains depend on the primary producers – plants that produce sugars via photosynthesis. Biomass is the mass of new biological material produced in a given area in a given time. An area with a high yield of biomass is very productive. Before I can explain biocapacity, I need to define biologically productive area.

> Biologically productive area is land and sea with significant photosynthetic activity and production of biomass. Marginal areas with patchy vegetation and non-productive areas are not included.
>
> (WWF, 2005)

Less than a quarter of the Earth's surface is biologically productive in a way that is useful to humans. The remaining three-quarters includes deserts, ice caps and deep oceans, which have low levels of bioproductivity that is difficult to harvest.

Table 6 Actual areas of useful land and water available globally

Area type	Description	World area available in hectares
Arable land	This is the most productive land type, growing plant crops for food, animal feed, fibre (e.g. cotton) and oil (e.g. sunflower oil).	3.3 billion
Grazing land	This land supports animals which provide meat, milk, wool, leather and other products.	1.7 billion
Fishing area	This includes sea fishing and freshwater lakes and rivers. Does not include deep ocean.	0.8 billion
Forest	Areas in which trees dominate. Most of the remaining forest is rainforest (3.6 billion ha), which is nearly nine-tenths of all the forest on the planet.	5.1 billion
Built-up land	This includes housing, roads, industrial areas and any other land where construction work has made substantial alterations (e.g. lakes and dams for hydroelectricity).	0.4 billion
	Total useful area available on Earth (less than one-quarter of surface area), assuming that all available area is used for human activities (i.e. no productive land or water is put aside for wildlife or wilderness)	11.3 billion

Source: WWF, 2005

This information is presented as a pie chart in Figure 57. In this type of diagram, the total area of the circle is divided into sectors. The larger the sector, the bigger the contribution from that particular item, for example forest. This chart shows the entire area of the Earth, including the useful areas shown in Table 6.

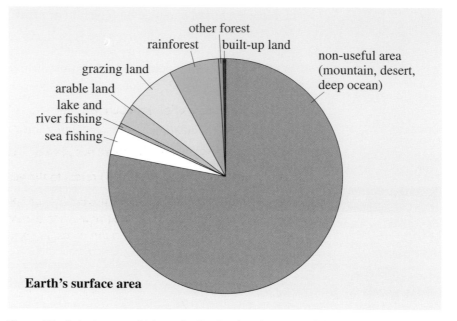

Figure 57   Actual areas of bioproductive land and water available globally

Source: derived from WWF, 2005

Activity 49

Allow about 15 minutes   Interpreting a pie chart

Compare Figure 57 with Table 6. Which do you find easiest to interpret?

Find the answers to the following questions:

(a) Roughly what proportion of the Earth's surface is useful for human activities?

(b) Of all the useful land types, which makes the largest contribution?

(c) Of all the forest types shown in Figure 57, which type covers the smallest area?

Comment

When comparing amounts that add up to a whole, a pie chart is often easier to interpret than a list of numbers, because you can see the relative amounts very quickly. If you need to know the actual numbers, it is easier to use the values from the table, because small differences are difficult to indicate clearly on a pie chart.

(a) About a quarter – as you can see from the pie chart.

(b) From the pie chart, forests and grazing land appear to make up about the same area. From the table, the numbers show that the total forest area is greater.

(c) 'Other forest' accounts for the least of the Earth's forest area.

Global bioproductive land per person

If we divide the total productive land by the world population, we can work out the area available per person:

Total world area of bioproductive land	11.4 billion hectares
Total world population	6.2 billion people
So, productive area of biosphere per person	1.8 hectares

This information from WWF (2005) refers to the year 2001.

An area 100 metres by 180 metres has an area of 1.8 hectares, so we can imagine an average plot of land, equivalent to the average area per person, and divide it up according to the main types, as shown in Figure 58.

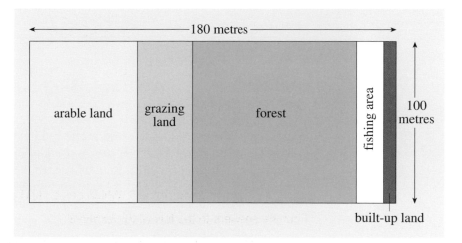

Figure 58   Global bioproductive area per person

This area is notional – an average over the entire world population – and there are many different land types gathered under these headings for simplicity. It does provide a useful measure against which to compare the use of resource in different continents and countries. Ideally, we should all be living within these limits, but what is actually happening?

> ## Biocapacity
>
> Biological capacity is a measure of how much resource is actually available to us in a given year. Although the area of useful bioproductive land is a good indication, we need to take into account the different productivities of various different land types in different areas. For example, a wheat field in the UK may produce more grain in a year than the same area of wheat in Zambia. Biocapacity is calculated by applying a conversion factor to each area of land, so a more productive wheat field effectively counts as a bigger area. The overall idea of adding up the contributions remains the same – it's just that each contribution matches the actual resources more realistically when biocapacity is used. Remember that biocapacity is a measure of what's available – so a bigger area is better (unlike ecofootprint, which is a measure of impact, so a smaller ecofootprint is better). Biocapacity is what's available – the more the merrier – while ecofootprint is what we use – less is best.
>
> > Biocapacity (biological capacity) is the total usable biological production capacity in a given year of a biologically productive area, for example, a country. It can be expressed in global hectares.
>
> (WWF, 2005)

7.3 Global resources – using information from charts and graphs

The total ecofootprint for the world is affected by the size of the population as well as the resource usage per person. Both of these are increasing. In the past, industrialisation and urbanisation have led to sharp increases in resource use per person, and that pattern is likely to continue unless alternative strategies are used. In this section, graphs and charts will be used to show the overall trends for various regions of the world, so you can improve your study skills as you work through it.

Figure 59 shows the total ecological footprint for the whole world (not per person) at specific times over the last 40 years, as a bar chart. In a bar chart, the length of each bar represents the number in that category. Each bar has the same width, and since the bars represent specific amounts at a given time, the bars do not touch so there are gaps between them. For any bar chart, you can interpret the information only for the values where there are bars – in this case, you do not know anything about the variation in between the dates given. Bar charts can be vertical, as in Figure 59, or horizontal. Some bar charts have divisions within each bar, showing how the different contributions add up to the total for that particular category. In this case, the key at the side of the chart shows the meaning of each shaded area. Where the chart shows a variation over a period of time, the time is usually marked along the horizontal axis, with the most recent times on the right.

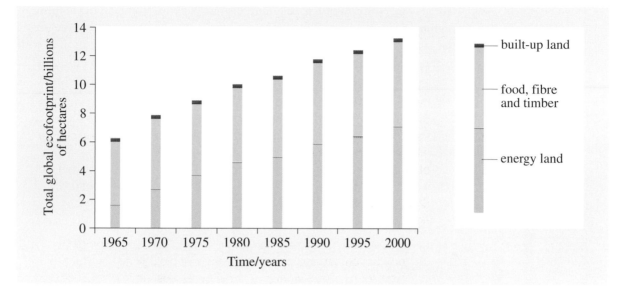

Figure 59   Changes in total global ecofootprint, showing the main contributing factors

Source: adapted from WWF, 2005, Figure 4

Activity 50

Allow about 10 minutes   How has the ecofootprint changed?

Use Figure 59 and information from earlier in this chapter (especially Table 6) to answer the following questions. Bear in mind that the vertical scale is in billions of hectares, so this is about different types of area contributing to the ecofootprint.

(a) What is the major factor contributing to the increase in ecofootprint?

(b) Using the categories defined in Table 6, what types of useful bioproductive area are included in 'food, fibre and timber', given that the entire ecofootprint is represented here?

Comment

(a) The contribution marked 'energy land' in Figure 59 shows the greatest increase. This is the 'energy land' discussed earlier in the book, indicating that the use of fossil fuels has been a major factor.

(b) There are five area types described in Table 6. One of these, built-up land, is marked separately in Figure 59, so the other four (arable land, grazing land, fishing area and forest) must be included in 'food, fibre and timber'. This seems to fit the description.

The ecofootprint for various regions

The resource use per person varies from one region to another. Generally, **developed** regions that have long-established industries, such as Western Europe and North America, use more resources per person than **developing** regions, such as Africa. Figure 60 shows the general pattern of resource use per person. Remember that this includes the 'energy land' contribution as well as the actual land types for food, timber and building. Assume that the global available biocapacity per person is 1.8 hectares. In Figures 60, 61 and 62, 'European Union' refers to the 25 member nations in 2005.

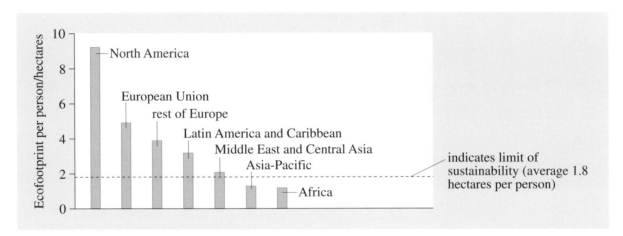

Figure 60   Ecofootprint per person for various regions   Source: adapted from WWF, 2005, Figure 5

The resource use per person is not the whole story: if a region has a large population, then the overall ecofootprint will be higher. Figure 61 shows the pattern of population in these regions.

Multiplying the ecofootprint per person by the number of people gives the overall ecofootprint for each region. Figure 62 shows the ecofootprint as an area – the larger the area, the larger the ecofootprint. To make sense of this, you need to know that each area is drawn to the same scale – the key at the top shows this. The dashed line shows the sustainable ecofootprint for each region.

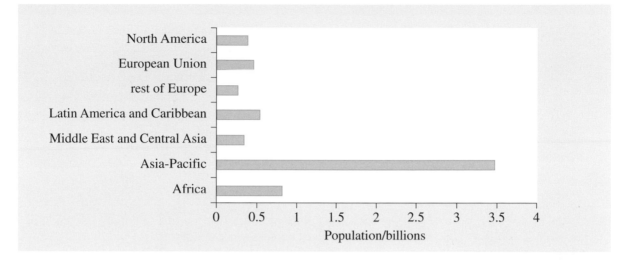

Figure 61 Population of various regions   Source: adapted from WWF, 2005, Figure 5

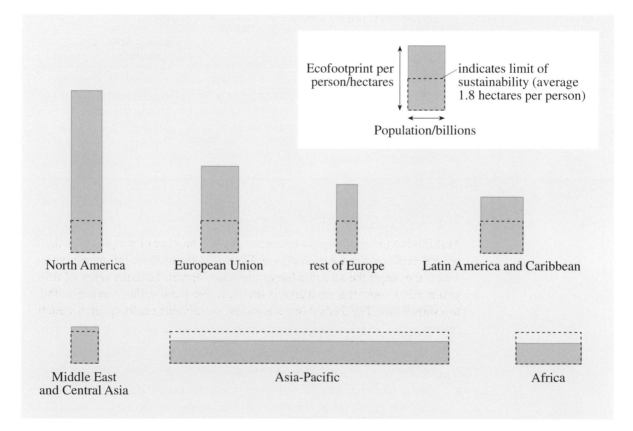

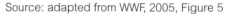

Figure 62 Overall ecofootprint of various regions   Source: adapted from WWF, 2005, Figure 5

Allow about 15 minutes

Comparing ecofootprints for different regions

From Figures 60, 61 and 62, answer the following questions:

(a) Which three regions have the highest ecofootprint per person?

(b) Which regions have the lowest ecofootprint per person, below the available biocapacity per person (i.e. sustainable)?

(c) Which region has the highest population?

(d) Which region has the lowest population?

(e) Which two regions have the highest overall ecofootprint?

Comment

(a) North America, the European Union and the rest of Europe are the three regions with the highest ecofootprint per person.

(b) Asia-Pacific and Africa have the lowest ecofootprint per person, below the available biocapacity of 1.8 hectares per person, so they are currently within sustainable limits.

(c) Asia-Pacific has the highest population.

(d) The 'rest of Europe' region has the lowest population.

(e) North America and Asia-Pacific are the two regions with the highest overall ecofootprint, but for different reasons. North America has a high resource use per person, with a relatively small population, but Asia-Pacific has a very large population (this region includes China), with a far lower resource use per person, at present. Adding together the two regions of Europe (within the European Union and outside it) makes a total ecofootprint that is similar to the resource use by either North America or Asia-Pacific.

7.4 Sustainability – trends in the UK and Europe

Figure 63 shows the variation in the UK ecofootprint over the last 40 years. This is an example of a line graph – a single line showing how one amount varies against another. As with any graph or chart, you should look for some standard information:

- The title of the graph describes what the graph is about.

- The horizontal axis and vertical axis (plural: axes) should be clearly labelled, so you can tell what measurements or amounts are being used.

- Where measurements are involved, the units should be stated – in this case, the ecofootprint is measured in hectares, and the time is identified by year.

- The source of the data should be given. Always consider whether it is a reliable source – is it likely to be fair, or does it represent only one point of view?

Figure 63 is typical of a graph that shows a variation over time. The year is plotted on the horizontal axis, with more recent times on the right. So, we say that the ecofootprint is plotted against time (the vertical axis is always plotted against the horizontal). Always check to see where each axis begins. In this case, zero is clearly shown for the vertical axis.

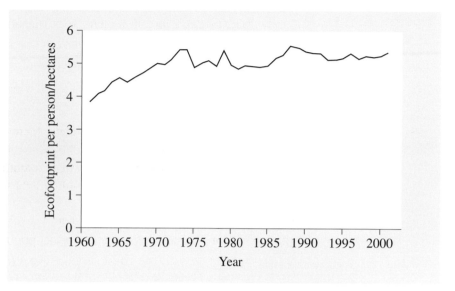

Figure 63   The UK ecofootprint per person

Source: adapted from WWF, 2005, Figure 16

Once you have checked the basic features of the graph, you can look at the overall shape, the trend. See whether the line slopes upwards or downwards. Look for any peaks (high points) or troughs (low points). The UK ecofootprint per person is similar to the value for other European countries, about 5 hectares per person. From the graph in Figure 63, there was a steady rise in ecofootprint during the 1960s and 1970s. Possible causes include increased car ownership alongside the greater use of consumer goods. Since then, the ecofootprint has remained steady, but at a relatively high level compared with 1961.

Whenever you are interpreting a trend from a graph, it is vital to read the scales carefully, so you know how big the variations are. Some graphs start at a value other than zero. Figure 64 (a, b, c and d) shows the same information as Figure 63, presented in four different ways.

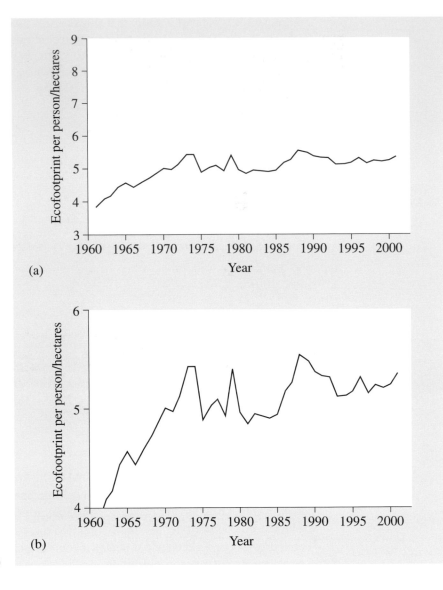

Figure 64 The UK ecofootprint plotted in different ways: (a) (b) (c) (d)

Allow about 15 minutes

Comparing graphs with different scales

Look at Figure 64 and write down how each graph differs from Figure 63. What is the effect of this on the way you interpret each graph?

Comment

(a) The scale starts at 3 hectares instead of zero – this makes the overall ecofootprint look smaller, because it is further down the scale.

(b) The scale starts at 4 hectares, and has been stretched, so the variations in ecofootprint have been emphasised – this makes the ups and downs in the graph look much more significant.

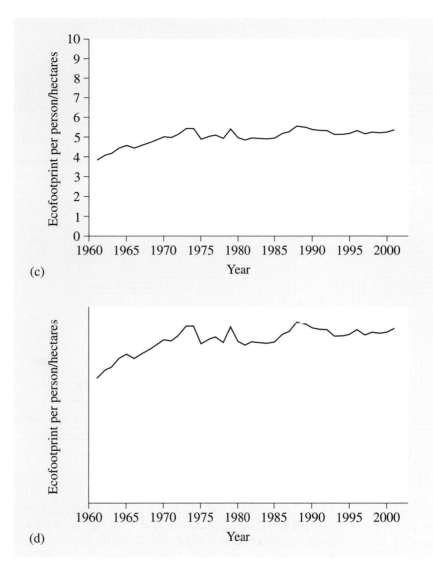

Figure 64   Continued

(c) The scale starts at zero, but goes up to 10 hectares, so the line has been squeezed into a narrower space. This makes the ecofootprint look smaller, but also makes the variations look less important.

(d) There is no scale on the vertical axis, although the rest of the graph is the same as Figure 63 – so this graph is difficult to interpret accurately.

Activity 53

Allow about 15 minutes   Ecofootprint trends in Europe

Using Figure 65, write brief notes describing the variation in ecofootprint per person over time in each of these countries. Which country appears to be the most successful in terms of use of resources?

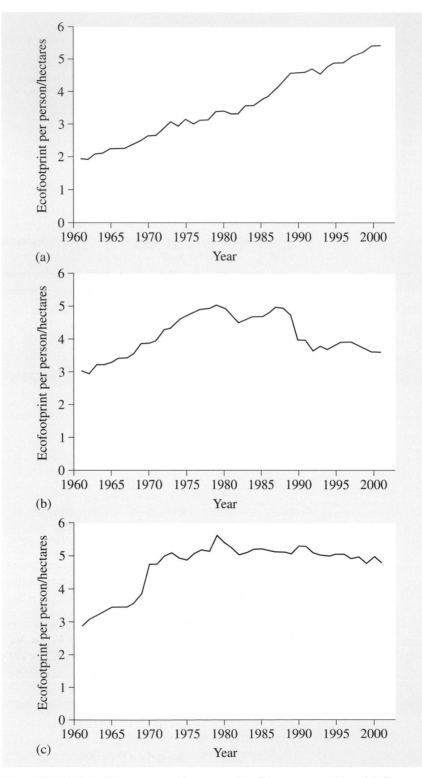

Figure 65 Ecofootprint per person for some other European countries: (a) Greece (b) Poland (c) Germany

Source: adapted from WWF, 2005, pp.8, 9

Comment

Greece: The ecofootprint per person is rising steadily.

Poland: The ecofootprint per person has risen, then, from 1985, the trend has been reversed and the ecofootprint has fallen. There is now a steady situation, as the ecofootprint line is now more or less horizontal, although still higher than the sustainable limit. This shows the most effective use of resources.

Germany: The ecofootprint per person has risen, then stabilised. The line has flattened off, and fallen slightly.

The graphs only show the changes – they do not indicate any reasons for the variation. Poland and Germany will be discussed later in this chapter. In the UK, possible reasons for the increase in ecofootprint include more car ownership, greater use of consumer goods, increased air travel and increased imports of food by air. For Greece, increased tourism (and air travel) and new buildings may have contributed.

7.5 Reducing the ecofootprint

According to the Worldwide Fund for Nature (WWF), there are four ways to reduce the overall ecofootprint:

- increase, or at least maintain, biocapacity
- lower the global population
- improve the resource efficiency with which goods and services are produced
- reduce the consumption of goods and services per person.

The last two strategies are the least difficult to implement in practice, although any reduction in the ecological footprint requires effort and investment.

From the graph in Figure 65(b), you can see that Poland has undergone a remarkable change since 1987. Poland's transformation is unusual, and makes an interesting story.

According to Cole (1995), the radical changes in Polish government after the end of the communist era, in the late 1980s, led to major shifts in environmental policy. Essentially, the economy collapsed, so financial reform was a major concern. There was widespread unemployment as factories closed down. There were also public protests about the severe air and water pollution left as a legacy of intensive industry, which was (and still is) causing widespread health problems. The explosion at the Chernobyl nuclear power plant in the neighbouring Soviet Union in 1986 further raised the profile of the environment on the political agenda. In 1988, large areas of the Polish countryside were protected as newly formed national or regional parks, covering 14 per cent of the country. Public awareness campaigns were started to encourage resource savings. In 1991, the government identified five ecological disaster areas, the result of inefficient and highly polluting industries. Financial penalties were established for any company exceeding the new pollution limits. Some industrial enterprises were threatened with closure if they did not comply. It was cheaper for Polish firms to reduce emissions than to pay environmental fines. State-owned electric power plants cut their emissions by 20 per cent in just two years, between 1989 and 1991. When the new Warsaw airport was completed in 1992, it was discovered to have avoided some of the legally required environmental protection measures. By this stage, the public outcry made headlines in the national press, and the new airport was closed until the necessary equipment was installed.

Poland's recent history is exceptional, and that country's environmental problems will take time to resolve, despite the progress made so far. Elsewhere in Europe, similar strategies have been tried. In the next section, I'll look at sustainable housing, then in Section 7.7, I'll consider waste management, to show how Germany has succeeded in making some positive changes. We'll also see what type of scheme would be needed to make a difference in the UK.

7.6 Sustainable housing – making a difference

'Sustainability' is a concept that can be understood in many different ways, depending on the circumstances. In this book, I'll take a relatively simple approach, although you may come across other definitions and interpretations if you read about sustainability elsewhere. For our purposes, a **sustainable** lifestyle or process is one that uses (or re-uses) the available resources, so that:

- overall, the resources used are within the limits of what is available (for example, if trees are used for wood, only part of the forest is cut down)
- there are other processes that balance the resource use (in our example, every time a tree is felled for timber, another one is planted)
- resource use can continue at that level over a long period of time, without depleting or removing the overall supply (wood is used slowly enough for there to be time for the new trees to grow)
- the use of resources is in proportion to their availability, so there is enough capacity to cope with unforeseen changes (enough of the forest has been left standing so that if there is a forest fire, the forest can recover through natural processes)
- the use of resources is effective (trees are cut down only when timber is actually needed, and wood is not wasted).

This may seem rather a long and complicated list, but it is, in effect, saying:

- take only what you need
- don't take more than is available
- leave enough to cope with unexpected events
- replace what you use.

Let's see how this might apply to housing.

Chapter 6 showed that a great deal of energy use (gas and electricity) is in people's homes. In the UK, domestic heating uses a large proportion of this energy. Cooking, heating water and running domestic appliances, such as refrigerators and televisions, also require fuel. Although, so far, I have focused on energy use, there are other environmental issues in the home, such as the use of water, particularly in the bathroom. The ecofootprint model does not include water supply, but, as you will see from the next track of the DVD, this is an important factor. Building new homes will also have an environmental impact from the materials used, and the energy to transport them and construct the houses. So, what can be done to make housing more sustainable?

Figure 66   The Hockerton Housing Project featured on Track 5 of the DVD

The four initiatives featured on Track 5 of the DVD are:

Social housing in Newark, showing how practical changes to existing homes can make a difference.

The Autonomous House, Southwell, a one-off house designed to have as low an environmental impact as possible.

Millennium Green, Collingham, a group of new homes designed to minimise environmental impact while appealing to more conventional tastes.

The Hockerton Housing Project, a low environmental impact community.

This video uses some specialised terms, so you may find the following definitions useful:

Super-insulation: extremely efficient insulation for roof, walls and floors, usually accompanied by double or triple glazing for the windows.

Passive solar gain: heat energy direct from the sun, absorbed by south-facing windows and walls of a building.

Envelope of a building: the ground floor, walls and roof.

Zero heating standards: a building built to zero heating standards does not need any heating from fossil fuels – solar heating and good insulation are sufficient.

Autonomous standards: similar to zero heating standards, but include collection and re-use of water, and composting of all biological waste, so the building is as self-contained as possible in terms of resources and waste disposal.

Solar panels: there are two types: some use the sun to heat water; others produce electricity (photovoltaic panels).

Photovoltaic cells (sometimes called PV cells): solar panels that convert sunlight directly into electricity.

Composting toilet: a toilet that does not use water or the usual plumbing. The waste is mixed with soil or other additives, and naturally decomposes into compost.

Low-E glass: glass that has been coated so that less heat is lost through windows (low emissivity glass).

Argon-filled double glazing: super-efficient double glazing, filled with the gas argon rather than air.

Earth shelter: house built into a bank of earth, often with earth as the roof, covered with grass.

Heat recovery unit: in a ventilation system, this unit exchanges heat between the incoming air and the outgoing air, so that as little heat as possible is lost (assuming that the aim is to keep the house warm).

Heat pump: a device that warms one place while cooling another, effectively moving heat from one area to another (a refrigerator has a heat pump that removes heat from the inside).

Thermal mass: an object or building with a large thermal mass will heat up and cool down slowly, evening out the effects of temperature variations outside.

Allow about 60 minutes

Greener homes

Making notes from video

This video lasts for about 45 minutes, so make sure you allow enough time to watch all of it. Of course, you can go back and have another look at sections if you find that watching it once is not enough. This is a documentary which presents a large amount of information. Just as with technical writing, it can take more time to understand the details.

As you watch the video, write down three things for each example that make these homes more eco-friendly (particularly for energy or water use). You may want to pause the DVD while you make notes.

Comment

See the comment at the back of the book.

Individual households not only use energy and water, but also produce large amounts of waste. Although each house may make only a small contribution to the overall amount of refuse, in a typical city there are so many households that the waste soon adds up. Industrial and commercial processes also produce waste. This is the subject of the next section.

7.7 Why does waste matter – and what can be done about it?

For York, which I will assume is a typical UK city, domestic waste accounted for 16 per cent (about one sixth) of the total ecological footprint in the year 2000. This is a significant proportion, so any scheme that reduces waste can make a useful difference. Recycling is often seen as a solution, but is this the best option?

In this section I'll consider the available options, and the implications for energy use, consumption of raw materials and demand for landfill sites.

To understand the choices available, we need to consider how a product is made, used and then disposed of. This is called the product life cycle.

Figure 68 illustrates this. For instance, the main raw ingredient of a glass bottle is silica (sand). Once the sand has been extracted from a quarry, it is purified and processed into glass by melting it at a high temperature, which requires a correspondingly high energy input. Bottles are then manufactured from the glass. In this example, assembly might mean filling the bottle with fruit squash, then distributing it to the supermarket.

Once the drink has been used, there are three options for the bottle:

- re-use the bottle, by returning it to the distributor to be refilled
- recycle the glass, by returning it to be melted down and manufactured into glass products
- dispose of the bottle. In the UK, this means burying it in landfill. As glass does not decay, it would simply remain there. Roman glassware has been found intact after nearly 2000 years of burial.

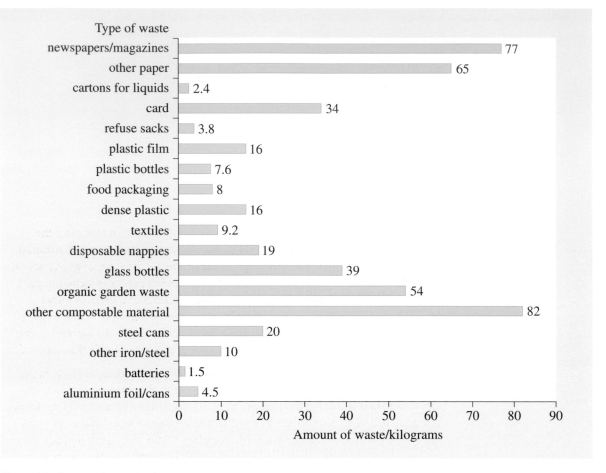

Figure 67 One year's worth of waste, averaged per person, for the city of York   Source: Barrett et al., 2002

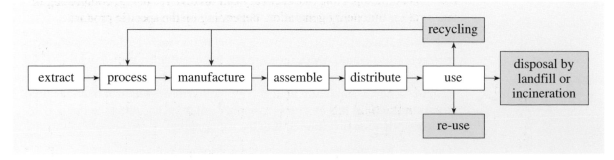

Figure 68 The product life cycle, showing the three options for post-consumer waste: re-use, recycling or disposal

---

Activity 55

Allow about 5 minutes   Waste choices

Of the three options (re-use, recycling, disposal):

(a) Which requires the most energy?

(b) Which has the least environmental impact?

(c) Which requires the least effort on the part of the producer and consumer?

Comment

See the comment at the back of the book.

These are the three options for **post-consumer waste** (i.e. waste after the product has been used). As you might guess, there may well be waste material and inefficient energy use at any stage in the product life cycle, from extraction onwards, so there are many possibilities for minimising environmental impact. In particular, there is a fourth option that is more effective than the other three. This is waste **reduction** – simply using less material, or less energy, at each stage. For the example of the glass bottle, this may mean increasing the efficiency of the glass kilns, or using more efficient transport in distribution.

Different products require different strategies. For glass bottles, refilling is the most effective, because it uses less energy than re-melting the glass to recycle it. For aluminium, a large amount of energy is needed to extract the metal from its ore, so recycling aluminium is worthwhile. It is also more difficult to make a refillable aluminium can than a refillable glass bottle. For many plastics, and cardboard, particularly when used in packaging, it would probably be best to reduce the amount used in the first place. Collecting, sorting and recycling these materials takes a lot of effort, because there are so many variations in the product. Even so, it is better to recycle them than dispose of them in landfill.

Figure 69 shows the priorities for waste management, with the most effective methods at the top. Note that recovery can involve recycling, composting or incineration for energy generation, depending on the specific product.

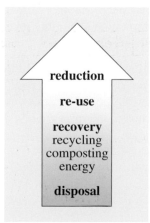

Figure 69 The waste management hierarchy, with the most effective methods at the top

This list of priorities applies to any product, from a bottle to a book to a car. The longer the item is used the better, because that makes best use of the energy that has gone into manufacturing it. Remember the discussion about carrier bags on Track 4 of the DVD – the shopping trip? Generally, disposable items have a larger environmental impact than items that are used for a long time. Re-use can mean, for example, taking unwanted clothes and shoes to a charity shop, or selling second-hand furniture, rather than dumping them in waste. For other items, re-use can mean mending or repair, rather than disposal at the first sign of a problem. Re-use is better than recycling, because it requires less energy, but recycling is in turn better than throwing items away.

Waste management in Germany

As you saw from the graph in Figure 65(c), Germany's ecofootprint has reduced since 1990. This is for various reasons. In this section, I will look at waste management, to show how government, corporate and individual action are all needed to make a difference. This section explains the history behind an eco-friendly idea that is now effective in Europe, and spreading to other regions.

By about 1991, Germany was producing surplus waste – there was no more space for landfill or other waste disposal, and the only option available was to export waste, an expensive process. In response, the German government, backed by a strong Green Party, put in place the 'ordnance on the avoidance of packaging waste'. This was the world's first **extended producer responsibility (EPR)** law. It made the manufacturers legally and financially responsible for the waste from their products. This law established the principle of 'producer pays' for the environmental consequences of a product or service. At the same time, waste recovery targets were set, for example 72 per cent of drinks containers to be refilled. Implementing this law required complicated and expensive changes for the various companies, and government systems such as waste collection and recycling plants. New national organisations were established – one to collect fees from businesses (products could then display the 'green dot' logo, showing that the producer had paid recycling fees in advance), another to recycle and process the waste. Public campaigns gave information about the 'green dot' logo, and recycling facilities were established (see Figure 70).

From 1991 to 1995, overall packaging decreased by 7 per cent in Germany (compared with a 13 per cent rise in the USA over the same period). In 1994, EPR was extended to other European countries via an EU directive, although each country was given scope to interpret the law differently.

(a)

(b)

Figure 70 (a) Pavement recycling bins in a German town – the labels are: Papier (paper); Restmüll (other rubbish); Glas (glass) (b) the 'green dot' logo

Extended producer responsibility (EPR)

The main principle is that 'the producer pays', so that anyone responsible for making or owning an item also has to dispose of it in an eco-friendly way.

• EPR extends the manufacturer's responsibility to the post-consumer stage of the product life cycle.

• Producers either take back or recycle their own products, or pay a third party to do so.

• Individual governments set targets for recycling, define what is to be recycled, and what is hazardous to use, and set targets for each stage of the process.

EPR has now been extended to larger, longer-term products such as electronic goods and cars. This can mean an increased cost, so the price of the item may increase, or there may be a charge for disposal (for refrigerators, for example).

Waste management in York

The principle of 'producer pays' has now become part of European law, so that many countries are taking steps to reduce waste. This is no easy task. The extra charges are not necessarily popular and new recycling facilities need to be built. For this to work, people and organisations need to support the schemes and recycle their waste – a change of culture.

In response to the 1994 EU directive, the UK government has set targets for the recovery of packaging waste and domestic waste recycling, among others. These targets were intended to be met over a period of time. For instance, in the year 2000, the latest targets were set:

- 25 per cent of household waste recycled or composted by the year 2005
- 30 per cent of household waste recycled or composted by the year 2010
- 33 per cent of household waste recycled or composted by the year 2015

This section will examine the impact of these targets on domestic waste disposal for York. What action would need to be taken to meet these targets, and how much would have to be done?

In the year 2000, in York, household waste made up roughly one-sixth of the overall ecofootprint – a large proportion. There were many components itemised as waste in the survey, but the largest contributions to the waste ecofootprint (including energy used in production) were due to:

Disposable nappies	12 per cent
Newspapers and magazines	19 per cent
Other paper and cardboard	29 per cent
Plastics	9 per cent
Textiles	9 per cent
Glass bottles	8 per cent

Activity 56

Allow about 10 minutes

Comparing numbers

The numbers in the UK waste disposal targets are presented as percentages, effectively an amount 'out of a hundred'. So, 30 per cent is 30 out of a hundred, which is the same as 3 out of 10, or three-tenths. Use the conversion table in Appendix 2 to look up the following percentages, and write them as fractions:

(a) 25 per cent

(b) 33 per cent

Use the conversion table in Appendix 2 if necessary to answer the following.

(c) List the six items from the York waste survey in order of their contribution to waste, largest first.

(d) Which item contributed more than a quarter of the total waste ecofootprint?

Comment

See the comment at the back of the book.

Several waste scenarios were considered, to see what effect these would have on the ecofootprint of household waste by the year 2010. A **scenario** is a plan for the future that considers the effects of decisions.

The first scenario, 'business as usual', would give a waste ecofootprint of 1.47 hectares per person by the year 2010, a significant rise. This assumed 11 per cent recycling, 15 per cent composting and a 3 per cent rise in the volume of waste each year, with the same population and the same types of waste.

The next scenario assumed that two-thirds of all York households recycled their waste, if a recycling scheme were introduced that collected from people's houses. All other assumptions remained the same. This gave an ecofootprint of 1.12 hectares per person by the year 2010, significantly lower than 'business as usual', but higher than the current value, so not reaching the sustainability target in the long run.

A more effective scenario included composting of all green waste (lawn mowings, kitchen waste and so on) by all households, as well as recycling other material. With this, the waste ecofootprint for 2010 could be reduced to 1.10 hectares per person, keeping it stable.

What would have to be done to actually reduce the ecofootprint?

The fourth, 'efficiency and sufficiency', scenario included recycling and composting, as well as significant waste reduction strategies. With all of this in place, the projected ecofootprint was reduced to 0.99 hectares per person, a genuine reduction. So what would have to be done to achieve this? This list gives some suggestions – to make the difference described here, many of them would need to be implemented by most of the residents of York (Barrett *et al.*, 2002):

- Don't use disposable carrier bags – take a re-usable bag to the shops.
- Purchase refillable containers wherever possible.
- Avoid using disposable products such as nappies, tissues, paper and plastic cups, razors.
- Avoid over-packaged goods and try to buy them unpackaged.
- Use rechargeable batteries rather than disposable ones.
- Use a milk delivery service (in glass bottles).
- Pass on unwanted clothes to friends or charity shops.
- Use and refill your own durable drinks bottle, rather than using throw-away plastic bottles.
- Discourage unwanted junk mail.
- Use spare parts for large electrical items, where possible, and get items repaired rather than disposing of them immediately when they break.

This chapter has covered a wide range of issues, extending the discussion to a global scale. You have seen that the overall ecofootprint for a region depends on the number of people (population) as well as the resource use per person. According to the ecofootprint model, our increasing use of resources, particularly energy, is taking us beyond sustainable limits. Sustainable living requires action at different levels: national (environmental legislation, taxes), corporate (changes in business policy, such as in supermarkets and building companies), local (recycling within a city) and personal (consumer choice). Even small personal changes can make a difference, if many people choose to make them. Although recycling and other schemes can help, the clearest way forward is to reduce our use of resources, although that is not always as simple as it may seem. In the next chapter you'll see how one particular community has brought about radical changes, and you'll have a chance to bring together the themes from the rest of this book.

Study checklist

You should now understand that:

- Bioproductivity and biocapacity are measures of the resources available.
- Less than one quarter of the Earth's surface is biologically productive.
- Sustainability is about using resources without depleting the overall supply, and can continue in the same way over a long time, allowing for unexpected circumstances.
- Comparing the total ecofootprint with the total biocapacity tells us whether current activity is sustainable.
- Housing can be made more sustainable by reducing energy and water use.
- Waste management (and waste reduction) can make a significant difference to the ecological footprint.

You should now be able to:

- summarise information
- develop an information-based argument
- interpret trends in graphs
- read information from tables, including units
- compare numbers in different forms, such as fractions and percentages, using a conversion table
- make notes from video.

8 What next?

8.1 Introduction

As the physicist Niels Bohr once said:

> Predictions can be very difficult – especially about the future.

As I wrote in Chapter 1, the environment is a vast subject, with many different strands and interconnections. Beyond this book, the debates will continue, and the issues may change as new information becomes available. The main principles will still apply – science and technology are constantly advancing, but each new idea or innovation is based on previous knowledge. That knowledge may itself be questioned and challenged, as part of the process of learning about the world. There will always be differing opinions, and different perspectives on the world. Science does not claim to answer every question – but scientific knowledge has the benefit of having been thoroughly tested in practical situations by many different people. Both science and technology can solve practical problems, but can also cause them, depending on how they are applied. Technology includes skills, social knowledge and ways of organising people, and as our awareness increases, these innovations can be applied in different ways.

8.2 Bringing the themes together

This chapter is not a conclusion, because the environmental story will continue, in newspapers, on television, in other books and study courses. It does offer you a chance to return to the themes of the course, and see how they apply in a practical situation. The first few chapters of the book introduced you to ecology, the branch of biology that covers the interrelationships between living things and their environment. This provided background information to explore the effects of human technological development, and some of the issues that raises.

Chapter 1 introduced the four themes that run through the book:

- timescales
- chains and cycles
- local and global
- diversity and sustainability.

Taking time to think about what you have learned is important – looking back, to reflect on your experiences. At this point, it is worth pausing to do just that.

Allow about 20 minutes

Reflecting on the themes

For each of the four themes, write a few notes (or draw a spray diagram if you prefer) to record what you can remember, or what seemed important, from the book so far. You do not need to go back through the book in detail, if at all – this is about your overall impression. If you have difficulty remembering the issues, look back through the contents list or the main chapter headings to remind you.

Comment

Whatever you have written, this is a useful technique to help you make sense of the overall pattern of ideas. Links between ideas, so long as they are not misleading, can be a very helpful way to remember what you have learned. For example, in the 'chains and cycles' theme, you may have linked the natural recycling of the carbon cycle with the need for people to recycle their waste. You can use this technique to make sense of large amounts of information, so you avoid getting stuck in the details – and it doesn't have to take long. Sometimes just spending five minutes jotting down your thoughts can be helpful.

Throughout this book, the text has been presented in relatively short sections, with activities and diagrams to help break it up into easily readable chunks. Elsewhere, information may not be presented in that way, so the next section gives you some practice in reading a longer account.

8.3 A case study – sustainable development

A **case study** is an extended example, intended to give a more in-depth account of a particular situation. Earlier in the book, the accounts of grouse management on Langholm Moor, and the information about the city of York, would both be classed as case studies. In this section, the case study takes the form of a fairly long article, written by a journalist, about a sustainable development in South Africa. The aim is to link various themes and ideas from this book to a real-world example.

Allow about 40 minutes

Relating ideas and themes to new information

Read the article 'An eco-town takes root'. As you read, make a brief note of any ideas or themes that you have encountered elsewhere in this book, or in the DVD audio or video. When you have finished, look up the comment at the back of the book.

Figure 71   Ivory Park ecovillage

An 'eco-town' takes root in South Africa

A poor community on the outskirts of Johannesburg paints its future green.

By the year 2007, for the first time in history, humanity will be primarily an urban species – *homo sapiens urbanis*.

Over 75 per cent of all North Americans and Europeans already live in cities, and in a few years most Southern citizens will have joined them. Each week another million people either move to or are born into a city. By 2015, there will be at least 23 'mega-cities' in the developing world, each with more than 10 million people.

Can the planet carry the weight of this urban sprawl? The world's cities take up just two per cent of the Earth's surface, yet account for 78 per cent of the carbon emissions from human activities – the biggest source of the greenhouse gases that contribute to global warming, says the Worldwatch Institute, a Washington, DC-based research institute.

Many cities also fail to provide decent living conditions for all their residents. Some 600 million to one billion urbanites lack adequate shelter and live without easy access to clean water, toilets or electricity.

The dilemma, however, is that for everyone to live at the same level of material wealth as North American city-dwellers, we'd need another three or four planets. Since good planets are hard to find, cities and towns have to better balance the demands of people with nature. They must leave smaller 'ecological footprints' by using less water and generating less waste, boosting self-reliance in food and energy, and promoting sustainable transportation.

Perhaps the most unlikely place for such a green town to take root is in the brown dirt of a slum called Ivory Park in South Africa.

Ivory Park, home to 200,000 people living in shacks topped with corrugated tin roofs, lies within the town of Midrand on the outskirts of Johannesburg. It is also home to grinding poverty. As many as 50 per cent of the adults are unemployed. Children play on potholed streets, dodging streams of polluted water. A permanent cloud of hazy smoke from tin-drum coal fires used for cooking hangs heavy. Respiratory illnesses are common.

Emerging from the ruddy soil is hope – in the form of brightly coloured buildings that make up the Ivory Park EcoCity village. Here, poverty eradication and sustainable development meet in a fusion of African and Western ideas to improve life for people and the planet.

At the edge of town is a market where farmers from six co-ops sell their organic produce. Within the village, several types of environmentally friendly houses are being showcased. The community centre, used for workshops and training, is a good example of the ecological construction common in Ivory Park. The centre is a round building of clay and concrete, with used polystyrene blocks as insulation and doors salvaged from a condemned building. It has a soil roof, with grass growing on the top and sloping sides so that children can play on it. The temperature inside is always comfortable, no matter what the season, because soil has insulating properties, keeping the centre warm in winter and cool in summer.

'It's just like having air conditioning,' says Annie Sugrue, an EcoCity managing trustee.

The Ubuhle Bemvelo Eco-Construction Co-op is building 30 of the eco-houses. The 14 women working in the co-op use indigenous materials and environmental building techniques adapted to local conditions. For example, thick earthen walls absorb heat during the day and radiate it during the cool nights of winters.

The houses are purchased through a housing subsidy process, with preference given to those involved in the EcoCity. 'We want people to live and work in the same place, as it cuts down on transport and pollution,' explains Sugrue.

The EcoCity was born as an experiment in alleviating poverty – and doing it without jeopardizing long-term ecological health. A non-governmental organization called Earthlife Africa first obtained a US$1.7-million grant from the Danish government in 1999. Since then, there has been funding from Canada, Switzerland and Sweden, and partnerships have been struck with the World Wildlife Fund and the United Nations Development Programme. The project is now 'owned' by the Johannesburg city council, and local and foreign businesses have come to the table, as have different South African government agencies.

The project is both a showcase demonstration and a training centre, but above all it's home for its residents. Mundane but pressing facets of life must be tackled.

For instance, to cope with the endless smoke from cooking fires, an energy centre encourages people to buy cleaner liquid propane in canisters and alternative energy equipment such as solar cooking ovens. One vital development has been the so-called 'smokeless umbhawula', an innovative tin-drum coal cooker that uses far less fuel than normal, and radically reduces the amount of unhealthy smoke.

Trash is another major problem in South Africa, but a successful Ivory Park waste recycling co-operative has offered an option. Now employing more than 40 people, bottles, glass, paper, plastics and tin are brought by waste collectors to a buy-back centre, which then sorts the waste and sells it to recycling companies.

A recycling project of a different kind is housed in a big metal shipping container. The Shova Lula (easy pedal) cycle co-operative imports second-hand bikes and parts from England, Germany and Switzerland, then bicycle mechanics repair and sell the working wheels to community members.

Involving youth has been key to the accomplishments of Ivory Park. Youth work at the bicycle co-op, serve as EcoCity guides and conduct environmental education workshops in schools and the community. A new project will train one hundred youth in eco-building techniques using earth bricks and passive thermal designs. Others will learn organic landscaping and farming techniques, and how to build biogas digesters that convert food and animal wastes into a clean fuel for cooking.

'The focus of the project is on self reliance,' says Annie Sugrue. 'South Africa has serious economic problems and at present cannot provide all the social services people need like electricity, clean water and sanitation. The poor feel they have little power to change things so they sit around and hope someone will come along and save them.'

'That's not going to happen.'

But community members do realize that a village needs to be sustainable and self-reliant – these are traditional African values. 'The EcoCity concept has been very well received by local people,' Sugrue says. However, the project is still heavily dependent on outside support.

Despite this, Ivory Park's ecovillage is expanding. More homes are being built, and initiatives such as large-scale biofuel production are on the horizon. Just as important, tens of thousands of people have visited the village, particularly during the World Summit on Sustainable Development in 2002, held in Johannesburg.

And good ideas travel: ecovillages are growing in South Africa and elsewhere.

In Senegal, for instance, 12 villages are being transformed into greener communities. Governed through community-based decision-making, the hearts of these villages beat with organic agriculture, solar energy and micro-enterprise economic development.

Internationally, an informal collection called the Global Ecovillage Network encompasses hundreds of small communities in Europe, North America and Asia. Some of these started in the 1960s as communes, and all are pursuing the objective of creating a better quality of life while living lighter on the planet.

At the larger end of the scale, many cities are seeking a greener future as well. The International Council for Local Environmental Initiatives, with over 2,000 member municipalities, is active in helping cities become more sustainable.

One member, Melbourne, Australia, home to 3.4 million people, is on target to eliminate its contribution of greenhouse gases by 2020. With comprehensive energy reductions of 50 per cent, the use of renewable energy and the absorption of local emissions in widely sowed native vegetation, Melbourne will likely be the first industrial city to achieve zero-net emissions.

And who knows what on earth Ivory Park will have accomplished by then.

The ecological footprint

Imagine if a giant glass bowl was inverted over the top of a city like Montreal, Canada. Nothing but sunlight could get in. Obviously most of the people inside wouldn't survive long because the city needs food, air, places to put waste and ways to obtain other resources from outside the city limits to sustain itself.

So how much bigger would the glass bowl have to be in order for everyone in the city to survive at their present standard of living? In other words, how big is Montreal's Ecological Footprint?

William Rees at the University of British Columbia has calculated ecological footprints for a number of hot spots around the world and determined that the average North American citizen uses the natural resources from 7.7 hectares (19 acres) of land.

And so Montreal with a population of 3.5 million has a whopping footprint of 270,000 sq. km, about half the size of France, and considerably more than the 462 sq. km the city actually occupies. Similar sized cities in the US and Australia have even bigger, more unsustainable footprints.

Because material resource use is lower in the Developing World – India averages just 0.5 hectare per person – those cities have much smaller ecological impact.

Source: Leahy, 2004

8.4 Writing – planning and making an argument

When you do your own writing, you will often have a task or a title in mind, such as the questions in Chapter 4 about rabbits and ecosystems. For longer pieces of writing, planning and structuring the text makes a big difference to the quality of the finished account.

Planning your writing is important, so that you can present your points in an order that makes sense to the reader. Think about the structure – what's the most important point? Do you need to describe some background information before you discuss the question? What order will make your points clear? There are often different ways to explain the same issue, so this is about personal preference. What matters is that what you write has to make sense to someone who does not know about the topic. A common fault is to miss out the basic information, because you've thought about the subject so much that it seems obvious by the time you start writing. This may not be the case for whoever is reading your work.

A paragraph is a group of sentences that cover related points – an important way to organise what you write. The pause between paragraphs gives the reader a chance to think about the meaning of each section. Text that is presented in one long page without paragraphs is very difficult to read. Equally, very short paragraphs can be disjointed and give the impression that the text is simply a long list of items. Two tips should help:

- Paragraphs should group sentences that relate to the same subject. The start of a new paragraph should coincide with a change of emphasis.
- You should reveal in your first sentence what a particular paragraph is going to be about. It provides a sense of direction and order to what you write.

The first paragraph of your account should set out the overall themes or issues, or at least highlight the main topic, so the reader knows what subject you are discussing. The following paragraphs can then set out each of your main points in turn. The last paragraph often wraps up the whole account with some sort of overall conclusion or closing thought. So, most accounts have a beginning, a middle and an end.

To organise your work, write a brief plan before you start. This can be as simple as a list of topics (one main one per paragraph). Then look at your plan – does it address the question? Have you covered all the points you want to make? Does the order make sense? At this stage, you can also check that you have all the information you need, and change the plan to fill any gaps.

As with any other skill, good writing takes time to develop, and even those of us who have been doing it for years still struggle at times. I find that if my plan doesn't make sense, or I keep changing the order of what I'm doing as I write, it can mean that I haven't had enough time to think about the subject. What I'm doing is making sense of my thoughts as I write them down. This is fine – in fact, it is a very good way to learn. It does become stressful at times, if I haven't allowed enough time before a deadline for this process. So, it is worth writing some notes, and forming a plan, well before the scheduled time for the finished work. This gives you some thinking time. If you are finding it difficult to plan, then just start writing – sometimes getting a paragraph, or even one sentence, on to paper can start your thoughts flowing. You can always change it later.

As an example, here is a version of my plan for the first few sections of Chapter 7 of this book, which is about 8000 words long in total. This is more complicated than a typical account, because it also includes the activities, but it shows the main points.

Chapter 7   Sustainability and our environment

7.1   Introduction

Ref. back to previous chapter – ecofootprint

New ideas – biocapacity, sustainability

Examples – reduce ecofootprint

7.2   How much useful land and water area is actually available?

Define biologically productive area

How much area? Global?

Compare types – ref. back to ecofootprint

Activity – pie charts

Box on biocapacity – not the same as bio. prod. area

7.2   Global resources – using information from charts and graphs

Intro – resource use – how much per person? How many people (population)? Ref. to Chapter 5

Graphs, charts – study skills

The ecofootprint for various regions (subheading)

Compare gobal regions – explain developed, developing countries

Resource per person, and population? (repeat – cut from previous section?)

Activity – comparing resources used by different regions – highest, lowest

7.3   Sustainability – trends in the UK and Europe

Changes in ecofootprint – use WWF graphs for Europe – example plus activity – trends

Define line graph, explain parts of graph

(misleading scales – extra activity?)

Looking for trends – activity

Graphs don't show why there have been changes – suggest factors

This is a fairly rough plan. In fact, every chapter in this book has been through at least three drafts, so the content changed at some stages. Even so, you can see where I have used abbreviations, such as 'bio. prod. area', and added queries – do I need an extra activity? Do I want to repeat an earlier point with a different example? Often, a plan is a private note, so you can write what suits you. Sometimes, you may need to show the plan or outline to someone else, for example if you are working on an assignment or a report. Planning matters – find a way that works for you.

One way to think about the structure of a piece of writing is to make notes from a published account. Of course, if you use or quote this information in your own work, you must include a proper reference. Even if you don't use the information, it is worth looking at published writing in more detail, to improve your own style. Reputable newspaper articles are written to make the main points as clearly as possible. Although not a scientific account, the Leahy article in this chapter provides an example. The technique in the next activity can be applied to any piece of writing, so you could use it to analyse the structure of a more formal scientific account from elsewhere. This task should give you some insight into how another writer structures an account.

Activity 59

Allow about 40 minutes

The structure of a written account

Read the Leahy article again. This time, as you go through, make a note of one main point from each paragraph. There are some very short paragraphs, as this is a journalistic article, so make a point for each main paragraph or change of topic. Write each note on a separate line, to produce a list like the example for my structure of Chapter 7. When you have finished, read your list of notes – does this give a fair impression of the structure of the article? What do you notice?

Comment

These are my notes on the article. I combined some of the very short paragraphs, because they covered similar topics. Your notes will be different – but see whether we have picked out similar points.

By 2007 most people will live in cities
Cities – high carbon emissions, sometimes low living standards
Need smaller ecofootprint
Example – Ivory Park, Johannesburg
EcoCity – combined ideas from different countries
Community centre – soil roof
Eco-construction – local materials
Reduce poverty – international funding
Energy centre – new type of cooker
Recycling – trash, bicycles
Environmental education, self-reliance
Other ecovillages starting elsewhere – Senegal, North America, Europe, Australia
(Ecological footprint – explanation)

From the structure, you can see how the author starts with a simple piece of information – that most people now live in cities – and then discusses the environmental issues. Once he has given us this background information, he then focuses on Ivory Park, which is his main example. He describes the main features of Ivory Park, and the environmental decisions that have been made. Towards the end of the article, he moves back to more general issues, to show how this example is being applied elsewhere. There is an extra section at the end that explains the ecofootprint, because the writer has assumed that the reader would not know how this works. If this were a more formal account, this extra explanation would need to be included in the main part of the text.

8.5 Next steps

I'd like to finish the book with a quote from Marshall McLuhan:

> There is absolutely no inevitability provided there is a willingness to contemplate what is happening.

(McLuhan and Fiore, 1967)

Now that you have finished this book, I hope you feel more confident in using the ideas and information presented in it. As you have seen, 'environment' is a fascinating and complicated subject. The science and technology are woven together, as are local and global issues. This book does not claim to be comprehensive, nor to provide any neat answers, but it should have raised your awareness, and may perhaps form the basis for further debate or study. You have also developed many skills, which will be useful whether you decide to continue studying, follow your interests through further reading, or make some practical changes to your own lifestyle.

Good luck in whatever you decide to do!

Study checklist

You should now understand that:

- Sustainable living and development depend on a complicated range of factors.
- Environmental awareness and positive steps can work at different levels: individual, community, organisation, government, international.

You should now be able to:

- reflect on the main ideas in what you have learned
- make sense of a longer written account
- plan and structure a written account.

References

Barrett, J., Vallack, H., Jones A. and Haq, G. (2002) *A Material Flow Analysis and Ecofootprint of York: Technical Report*, Stockholm, Stockholm Environment Institute.

Cole, D.H. (1995) 'An invisible hand for Poland's environment', *Wall Street Journal – Europe*, 27 September. Available from: www.publicinternationallaw.org/publications/editorials/Invisible%20Hand.htm [accessed 22 September 2006].

Colinvaux, P. (1978) *Why Big, Fierce Animals Are Rare*, London, Allen & Unwin.

Darwin, C. (1985, first published 1859) *The Origin of Species*, London, Penguin Books.

Fortey, R. (1998) *Life: An Unauthorised Biography*, London, Flamingo.

Guardian (2005) 'The real cost of a £50 ticket from London to Edinburgh', 7 July. Available from: http://travel.guardian.co.uk/ecotourism/story/0,,1523118,00.html [accessed 12 August 2006].

Hecht, J. (1995) 'The geological timescale' (Inside Science, No. 81), *New Scientist*, Vol. 1978, 20 May.

Keeling, C.D. and Whorf, T.P. (2005) 'Atmospheric CO_2 records from sites in the SIO air sampling network' in *Trends: A Compendium of Data on Global Change*, Carbon Dioxide Information Analysis Center, Oak Ridge National Laboratory, US Department of Energy, Oak Ridge, TN.

Leahy, S. (2004) 'An eco-town takes root', www.sustainabletimes.ca/articles/EcoCity.htm [accessed 15 August 2006].

MacLulich, D.A. (1937) *Fluctuations in the Numbers of the Varying Hare (Lepus americanus)*, Toronto, University of Toronto Press.

McLuhan, M. and Fiore, Q. (1967) *The Medium is the Massage: An Inventory of Effects*, New York, Bantam.

Myers, N. and Kent, J. (eds) (2005) *The New Gaia Atlas of Planet Management*, London, Gaia Books.

Open University (1998) S103 *Discovering science*, Block 10 'Earth and life through time', Milton Keynes, The Open University.

The Times (1998) 'An inglorious ending for Churchill's grouse moor', *The Times*, 27 October, p.3.

Thomas, C.D. et al. (2004) 'Extinction risk from climate change', *Nature*, Vol. 427, 8 January, pp.145–8.

Williams, M. (2001) 'The history of deforestation', *History Today*, Vol. 51, No. 7, pp.30–7.

WWF (2005) *Europe 2005: The Ecological Footprint*, Brussels, WWF European Policy Office. Available from: http://assets.panda.org/downloads/europe2005ecologicalfootprint.pdf [accessed 12 August 2006].

Answers to activities

Chapter 2

Activity 9

Organisms:

- wide range of organisms (living things) including animals, plants, fungi, bacteria and so on
- scientific names – underlined if handwritten; Latin; genus and species

Environment:

- physical environment (not living)
- can be seen in terms of ecosystems

Habitat:

- where something lives

Interrelationship:

- many connections even in small rockpool
- for example, food chain, food web

Activity 12

Some of the other interrelationships are:

- caterpillars eat oak leaves
- robins eat caterpillars
- sparrowhawks eat robins
- humans eat a wide range of plants and animals.

Activity 14

(a) grass → snail → thrush → cat

(b) oak leaf → greenfly → ladybird → robin → sparrowhawk

(c) dandelion → rabbit → human being

(d) grass → snail → hedgehog → fox

Chapter 4

Activity 27

(a) The reasons for changes to forests and hedges:

Roman:

- new hedges to mark ownership boundaries
- wood used for fuel
- trees cleared to give view from roads

Norman and later:

- protected some forests for fallow deer (food and sport)
- demands for timber and fuel reduced forests
- hedges planted when fields enclosed

Victorian:

- no longer needed wood for charcoal so cut down old forest and replaced it with fast-growing pine trees or farmland
- new hedges resulting from changed patterns of land ownership

1950s:

- mechanised farming – tractors
- hedges removed to make larger fields

1990s:

- more hedges planted

(b) Other changes in land use:

Roman:

- new roads
- quarrying for clay and gravel
- growth of cities (built-up land)

Victorian:

- coal mines
- new canals
- more/improved roads
- first railways

Activity 28

Technologies before 1750:

Cultivating soil/farming:

- digging stick
- ox-/horse-drawn plough
- letting land lie fallow to replace nutrients
- applying animal manure to the land
- two-, three- and four-field rotation
- seed drill
- using clover to increase nitrogen in the soil
- selective livestock breeding

Economy:

- coins (Roman)

Travel:

–   Roman roads (straight, well drained)

–   boats

Building:

–   bricks

Activity 29

Technologies since 1750:

–   Using coal for fuel (steam engines, steel furnaces)

–   Canals for transport

–   Railways

–   Artificial nitrogen fertilisers

–   Tractors

–   New cereal crops: more effective use of nitrogen

Don't forget to write a paragraph comparing farming before 1750 with farming after 1750 – you can use your notes from Activities 28 and 29. I'm not providing a sample paragraph, because it's important that you try this for yourself, and save what you write as you'll need it later.

Activity 32

Your diagram may well differ from the one shown here, but this shows some of the interrelationships described in the text.

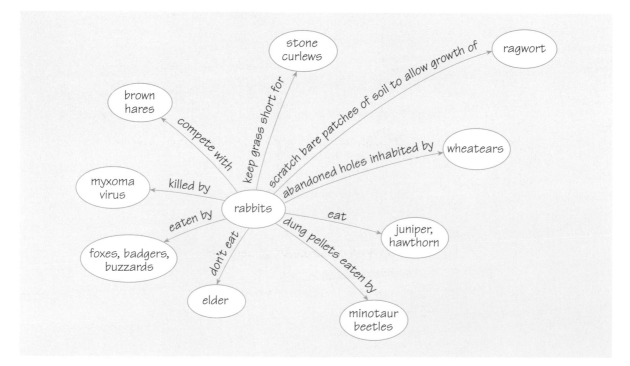

Figure 72 Interrelationships between rabbits and other organisms on downland

Chapter 5

Figure 73 shows the point at 20 weeks when the population is 33 mice.

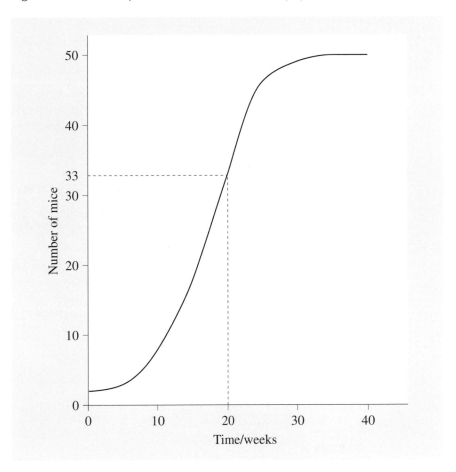

Figure 73 The population of mice over time, with a specific point marked

Chapter 6

(a) Arable land – the carrots are a plant crop

(b) Grazing land for the sheep

(c) Forest – paper is made from wood pulp

(d) Built-up land

The current level of carbon dioxide in the atmosphere is about 382 p.p.m., so the prediction made in 1990 was correct.

(a) 5 hectares is 5 × 10 000 square metres, which is 50 000 square metres

(b) 8 square kilometres is 8 × 100 hectares, which is 800 hectares

(c) 230 square kilometres is 230 × 1 000 000 square metres, which is
     230 000 000 square metres

Chapter 7

Activity 54

The first time I watched this video, I was left with one overall point: we can
make changes to existing houses, or build more eco-friendly houses, but
sometimes it is more about how we live, rather than the design of the house.

Here are some of the features I noted – there are many more on the video.

Example 1: social housing in Newark

1   Use the heating timer and temperature controls more carefully

2   Install double glazing

3   Heat each room separately, and turn off heaters not in use

Example 2: The Autonomous House, Southwell

1   Design houses with windows facing the sun, to absorb heat and allow in light

2   Use solar cells to generate electricity

3   Use a composting toilet to save water

Example 3: Millennium Green, Collingham

1   Design the house with extra insulation

2   Use solar panels to provide hot water

3   Use local materials, to reduce the impact of transport

Example 4: The Hockerton Housing Project

1   Build the house into the earth, for insulation

2   Install a wind turbine

3   Use a reed bed to process sewage

Activity 55

(a) Recycling, because the glass has to be transported and re-melted.

(b) Refilling, because, although the bottle has to be transported and cleaned,
     this takes less energy than melting glass.

(c) Probably disposal (throwing it in the bin), unless there is a local recycling
     scheme.

Activity 56

(a) One quarter ($\frac{1}{4}$) is the same as 25 per cent.

(b) One third ($\frac{1}{3}$) is the same as 33 per cent – or near enough!

(c) The six items in order of their contribution to waste, largest first, are:
     paper and cardboard, newspapers and magazines, nappies, plastic and
     textiles, glass bottles.

(d) Paper and cardboard contributed over a quarter of the total waste
     ecofootprint. One quarter is 25 per cent, which is less than the 29 per
     cent quoted.

Chapter 8

As ever, your response will be a personal one, so it will differ from mine. What matters is that whenever you learn new ideas, you are then able to apply them in a new situation. This may not happen easily, so be prepared to go back and think about what you have learned, and how it might apply. Once you have finished this book, you could continue to look out for news articles or other information, and see how the themes and ideas apply.

Here are a few of the thoughts I jotted down:

Cities – Chapter 4 – the industrial revolution – people in the UK started moving into cities, so what's happening now is similar to that.

Ecofootprint – Chapter 6 covered this.

Community centre – built with an earth roof and other insulation, from local materials – like the Hockerton housing project in the DVD video.

Trash recycling – the same problem as in Germany and the UK (Chapter 7).

Organic landscape and farming techniques – these can shape the local environment, because they change the use of the land (Chapter 4).

Appendix 1 The geological timescale

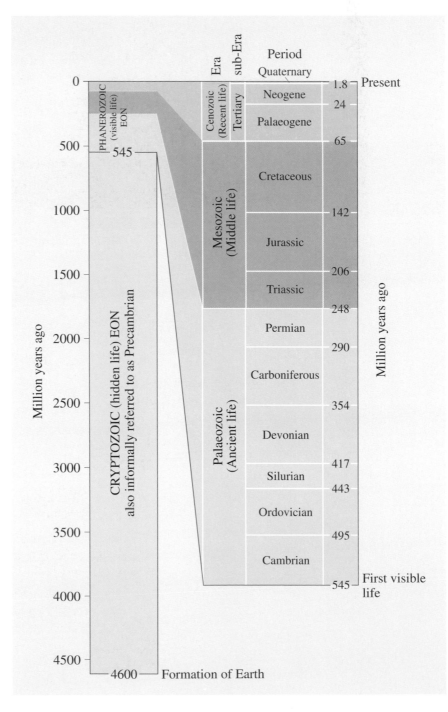

This geological timescale is provided for information only – you are not expected to remember any of it, although you may find it useful for reference. The earliest (oldest) events are at the bottom of the diagram, with the latest (youngest) at the top. The right-hand side of the diagram is an expanded version of the Phanerozoic Eon (pronounced 'fan-era-zoh-ic ee-on'). This covers events since 545 million years ago, when visible life first appeared. Before that was the Cryptozoic Eon ('krip-toh-zoh-ic ee-on'), the eon of hidden life. All the timescales are in millions of years.

Within the Phanerozoic Eon, standard Greek and Latin names are used by geologists to identify the eras and periods. Although many of these may be new to you, you may recognise the Jurassic Period, a time when dinosaurs were active. The Carboniferous Period (meaning 'carbon-carrying') was a time when large areas of the Earth were covered by rich vegetation, which eventually formed much of the coal available today.

Appendix 2 Conversion charts for fractions, decimals and percentages, and measurements of distance and area

Fraction	Decimal	Percentage
one hundredth 1/100	0.01	1%
two hundredths 2/100 or one fiftieth 1/50	0.02	2%
five hundredths 5/100 or one twentieth 1/20	0.05	5%
one tenth 1/10	0.1	10%
15/100	0.15	15%
two tenths 2/10 or one fifth 1/5	0.2	20%
one quarter 1/4	0.25	25%
three tenths 3/10	0.3	30%
one third 1/3	0.33 approx	33% approx
two fifths 2/5	0.4	40%
one half 1/2	0.5	50%
six tenths 6/10 or three fifths 3/5	0.6	60%
two thirds 2/3	0.66 approx	66% approx
seven tenths 7/10	0.7	70%
three quarters 3/4	0.75	75%
eight tenths 8/10 or four fifths 4/5	0.8	80%
nine tenths 9/10	0.9	90%
one	1	100%

Conversion chart for measurements of distance

Metres	Kilometres	UK old English units*
1	0.001	0.000 621 miles = 3.28 feet
10	0.01	0.006 21 miles = 32.8 feet
100	0.1	0.0621 miles = 328 feet
1000	1	0.621 miles
10 000	10	6.21 miles
100 000	100	62.1 miles
1 000 000	1000	621 miles

* to 3 significant figures

Conversion chart for measurements of area

Square metres	Hectares	Square kilometres	UK old English units*
1	0.0001 or one ten-thousandth	0.000 001 or 1/1 000 000	10.8 square feet
10	0.001 or one thousandth	0.000 01 or 1/100 000	108 square feet
100	0.01 or one hundredth	0.0001 or 1/10 000	1080 square feet = 0.0247 acres
1000	0.1 or one tenth	0.001 or 1/1000	0.247 acres
10 000	1	0.01 or 1/100	2.47 acres = 0.003 86 square miles
100 000	10	0.1 or 1/10	0.0386 square miles
1 000 000 or 1 million	100	1	0.386 square miles
10 000 000 or 10 million	1000	10	3.86 square miles
100 000 000 or 100 million	10 000	100	38.6 square miles
1 000 000 000 or 1000 million or 1 billion	100 000	1000	386 square miles

* to 3 significant figures

Source for conversion factors: National Weights and Measures Laboratory (2005) www.nwml.gov.uk/faqs.aspx?ID=8 [accessed 1 October 2006]

Acknowledgements

Grateful acknowledgement is made to the following sources for permission to reproduce material in this book.

Images

Cover image: Copyright © Rubberball/Getty Images; Figure 1: Copyright © European Space Agency/Science Photo Library; Figure 3: Plates 1–3, Matthews, S. and Missarzhevsky, V. (1975) Journal of the Geological Society, Vol. 131, Geological Society Publishing House; Figure 7: OAR/National Undersea Research Program (NURP)/NOAA; Figure 8: Photo by Woods Hole Oceanographic Institution; Figure 11a–c: Copyright © Photodisc #44/Getty Images; Figure 11d: Photos by Josje Snoek and Petra Visser (University of Amsterdam), compilation by Ellen Spanjaard; Figure 12a: © Natural Visions/Alamy; Figure 12b: Copyright © HAWKEYE/Alamy; Figure 12c: Copyright © Ann and Steve Toon/Alamy; Figure 12d: Copyright © Dr P. Marazzi/Science Photo Library; Figure 14: Copyright © Natural History Museum, London; Figure 21: Photo by Alice Peasgood; Figure 22: Photos by Alice Peasgood; Figure 24: Copyright © woodlandpictures/Tim Dent; Figure 28a: Photo by Alice Peasgood; Figure 28b: Copyright © Wildscape/Alamy; Figure 29: Photos by Alice Peasgood; Figure 34: Dembinski/FLPA; Figure 53a: Copyright © Transport for London; Figure 53b: Reproduced with the permission of Ordnance Survey on behalf of The Controller of Her Majesty's Stationery Office © Crown copyright. The Open University, Milton Keynes, Licence No.100018362; Figure 56a: Copyright © V K Guy Ltd; Figure 56b: Copyright © Michael Jenner/Alamy; Figure 66: Copyright © Hockerton Housing Project; Figure 70a: Photo by Alice Peasgood; Figure 70b: Copyright © Der Grüne Punkt – Duales System Deutschland GmbH; Figure 71: Copyright © Alex Webb/Magnum Photos.

Text

Pages 160–164: Leahy, S. (2004) 'An Eco-Town Takes Root', presented by the International Development & Environment Article Service, supported by CIDA.

Index